Optimization Methods in VLSI Design

Apoorva S. Shastri

Optimization Methods in VLSI Design

A Machine-Generated Literature Overview

Springer

Apoorva S. Shastri
Institute of Artificial Intelligence
Dr. Vishwanath Karad MIT World Peace University
Pune, Maharashtra, India

ISBN 978-981-95-2456-3 ISBN 978-981-95-2457-0 (eBook)
https://doi.org/10.1007/978-981-95-2457-0

This Springer imprint is published by the registered company Springer Nature Singapore Pte Ltd.
The registered company address is: 152 Beach Road, #21-01/04 Gateway East, Singapore 189721, Singapore

Preface

Traditional and modern electronics technology has always been at the forefront of every computer and automation-related innovation. The efficient utilization of the processing space is very much required for the millions of metal oxide semiconductors (MOS) to be fit on the smallest size of chips. The very large-scale integration (VLSI) design is at the core of this progress. Several academic researchers and industry practitioners have been working toward developing and applying heuristic methods for the past couple of decades. The vast literature associated has to be presented in quite an organized way. In light of this, this book reviews several algorithms and methods used for solving optimization problems in VLSI design. It introduces optimization methods in VLSI design using meta-heuristic algorithms and how they can be applied to problems like physical design, floor planning, etc. It provides a review of high-level synthesis techniques, such as measuring the quiescent current from the power supply, crosstalk noise mitigation methodology, and geometric programming for gate-sizing to reduce the design time of a VLSI circuit. This book delves into power grid synthesis, efficient testing and verification methods, optimization approaches for clocking, and delay minimization. It explores approximate computing techniques in VLSI design. Additionally, readers will understand how to achieve efficient full-chip optimization. It equips readers with the necessary knowledge and techniques to tackle design challenges, optimize performance, and achieve efficient and reliable VLSI designs in today's competitive landscape. This book is written for researchers, professionals, and students working in the core areas of electronics and their applications, especially in digital VLSI design and systems.

Furthermore, the review covers the rationale of the problem being addressed, the associated short literature supporting the contribution, the choice of optimization methods and approaches, the case studies, results, and the significance of the solutions. Such classified review may help UG and PG students, professors, and researchers from academia as well as industry gain a critical insight at the various systems. The articles chosen as well as the reviews are necessarily driven by the keywords recommended by the editor of the book as well as those generated by the machine based on the suggestions.

Many thanks to the Editorial Advisors, Shrutika Dasari, Chaitra Jawalkoti, Ha-Linh Vu, and Prajana Kate as well as Kamiya Khatter, the Associate Editor at Springer, for their unparalleled support in several ways. Their help was essential in conceptualizing and bringing this book to fruition. Thanks are also due to Shivam Silswal, MITWPU, for his valuable and timely help in organizing and finalizing this book.

Pune, Maharashtra, India Apoorva S. Shastri

About This Book

Auto-summaries can be generated by either abstractive or extractive auto-summarization:

- An extraction-based summarizer identifies the most important sentences of a text and uses the original sentences to create the summary.
- An abstraction-based summarizer creates new text based on deep learning. New phrases are created to summarize the content.

The auto-summaries you will find in this book have been generated via an extractive summarization approach.

Each chapter was carefully edited by Dr. Apoorva Shastri. The editors selected the papers, which were then auto-summarized. The editors have not edited the auto-summaries due to the extraction-based approach and have not changed the original sentences. You will find the editors' reviews and guidance on the auto-summaries in their chapter introductions.

In machine-generated books, editors are defined as those who curate the content for the book by selecting the papers to be auto-summarized and by organizing the output into a meaningful order. Next to the thoughtful curation of the papers, editors should guide the readers through the auto-summaries and make transparent why they selected the papers.

The goal is to provide a current literature review of Springer Nature publications on a given topic to support readers in overcoming information overload and to help them dive into a topic faster; to identify interdisciplinary overlaps; and to present papers which might not have been on the readers' radar.

Please note that the selected papers are not used to train an LLM while the auto-summaries are created.

Contents

Chapter 1
Overview of Optimization Methods in VLSI Design

Introduction by the Editor

VLSI design is the backbone of all tech breakthroughs happening around the world. Chip development has followed Moore's law. Through which chips have been able to pack more transistors in smaller area with less power consumption. Floor planning, a critical step in chip development which distributes design block on silicon, becomes more and more complex as additional transistors are placed with advance process node technologies. This requires a lot of optimizations to minimize the design block footprint, reduce interconnect delay between the same and leave least amount of dead space. This chapter talks about such optimization techniques including Genetic algorithms (GA) and Honey Badger algorithm (HBA). Main objective of this chapter is to compare the performance of various Nature-inspired methods in discrete optimization problems. A floorplan can be discretely viewed as sliceable and non-sliceable floorplan. A sliceable floorplan is achieved by recursively halving the floorplan area until each non-overlapping partition reaches a set defined size of a design block. Some design blocks don't allow a allow a floorplan to become sliceable. Such non-sliceable floorplan optimization is done with a Genetic algorithm called Lion optimization algorithm (LOA). LOA uses power B* tree crossover operator to solve a multi objective problem. It is inspired by the behaviour of a Lion and Lioness. Both Lion and Lioness from a pride come together and protect the young offsprings as well each other from predators. Implementation of LOA with B*tree crossover operator proves to be an efficient and effective technique for VLSI floorplan optimization. In this chapter, application of Simulated Annealing (SA) on combinatorial optimization is also discussed. This study aims to minimize dead space on the floorplan. SA is like the annealing process used in metallurgy. SA starts with highest degree of randomness (like high temperature in Annealing) as a function while exploring the solution space. We reduce the randomness by coming near to a solution (like low temperatures in Annealing). A solution which decreases

A. S. Shastri (ed.), *Optimization Methods in VLSI Design*,
https://doi.org/10.1007/978-981-95-2457-0_1

randomness of placement explores all possible states. Solution acceptance criteria is defined depending on the initial function value. The acceptance probability distribution decides whether to accept the new solution or to bounce out of it. To overcome an all-in exploration approach in traditional SA algorithm, authors have developed Simulated Spheroidizing Annealing Algorithm (SSAA). In this method, the time for local search is extended at any given incremental solution is increased. Furthermore, to improve the method of finding neighbourhood solutions, additional perturbation pattern approach is used until it reaches the desired iteration. This approach is termed as Improved Simulated Spheroidizing Annealing Algorithm (ISSAA). The results shows that the proposed ISSAA outperformed SSAA for larger circuits. However, perform is comparable while solving for larger die size or dimensions. In addition, by using parallel computation method an attempt is made to reduce the computational time, i.e., CPU time. Particle Swarm Optimization (PSO) based VLSI floorplan optimization is also discussed. A design block or Swarm in this algorithm, is placed in best location possible and its connecting blocks follow this swarm to accommodate itself in the best location in the same neighbourhood. PSO is weak with discrete exploration space and planar systems. Other algorithms can always be appended to get an efficient solution. Moving along, GA is population-based optimization technique. Population is updated after certain amount of time; a set of solutions can be procured with every update for GA to better search in each space. GA uses defined probability rules to fine tune the exploration. Fundamental remains same, converge population to local optimal solution. These principles of mutation and crossover are adopted and incorporated in PSO to achieve better diversity. Experimental results prove the proposed algorithm can achieve optimal solution for complex module placements and can be applied to multi-objective problems in floorplan. It also addresses problems in VLSI floor planning such as fixed-outline floorplan and Thermal-Aware Non-slicing floorplan. Additionally, the use of mathematics-based metaheuristics and hybrid machine learning clustering for discrete optimization problems in VLSI is also explored.

GA, PSO, HBA, LOA, SSAA and ISSAA are nature inspired algorithms. With fine-tune modelling of these algorithms, huge levels of optimizations can be achieved, and lot of churns can be avoided in floor planning.

Machine Generated Summaries

Disclaimer: The summaries in this chapter were generated from Springer Nature publications using extractive AI auto-summarization: An extraction-based summarizer aims to identify the most important sentences of a text using an algorithm and uses those original sentences to create the auto-summary (unlike generative AI). As the constituted sentences are machine selected, they may not fully reflect the body of the work, so we strongly advise that the original content is read and cited. The auto generated summaries were curated by the editor to meet Springer Nature publication standards. To cite this content, please refer to the original papers.

Machine generated keywords: floorplanning, floorplan, btree, module, floorplanning problem, VLSI floorplanning, dead, dead space, wirelength, representation, match, framework, optimization problem, floorplanne, interest

Comparing the Performances of Six Nature-Inspired Algorithms on a Real-World Discrete Optimization Problem [1]

This is a machine-generated summary of:

Hakli, Huseyin; Uguz, Harun; Ortacay, Zeynep: Comparing the performances of six nature-inspired algorithms on a real-world discrete optimization problem [1]

Published in: Soft Computing (2022)

Link to original: https://doi.org/10.1007/s00500-022-07466-1

If you want to cite the papers, please refer to the original.

For technical reasons we could not place the page where the original quote is coming from.

Abstract-Summary

"In order to guide and give insight into the practitioners and researchers about this problem, studies involving the comparison and evaluation of the performance of algorithms are needed."

"The performances of six nature-inspired methods, which included five new implementations of differential evolutionary algorithms (DE), scatter search (SS), equilibrium optimizer (EO), marine predators algorithm (MPA), and honey badger algorithm (HBA) applied to land redistribution problem and genetic algorithms (GA), were compared."

"In order to compare the algorithms in detail, various performance indicators were used as problem based and algorithm based."

"Experimental results showed that DE and SS algorithms have a more successful performance than the other methods by solution quality, robustness, and many problem-based indicators."

Introduction

"The two primary aims of the study are to compare the performances among different nature-inspired algorithms on discrete optimization problems and improve the solutions for the land redistribution problem with the five new implementations."

"Sections 1.1 and 1.2 present a literature review comparing the algorithms' performance for real-world problems and on the land redistribution problem."

"Many studies compare the performances of many nature-inspired algorithms implemented for a real-world problem in published literature."

"Of investigating the literature review on the land redistribution problem and comparing the algorithms' performance to the real-world problem, the following issues have arisen as the motivation of this study."

"Despite the previous studies, performance evaluation of these algorithms when applied to land redistribution problems is not shared."

"SS and DE based on evolutionary algorithms, MPA and HBA based on swarm intelligence, and EO based on physics are applied to land redistribution problems."

Material and Methods

"Before the land redistribution process, blocks are created in the block planning phase."

"The primary aim of the land redistribution is to allocate the parcels into the blocks, which are the preferences of landowners, by providing the minimum overflow/residual areas for blocks."

"More or less area occurs in the blocks due to the land redistribution process."

"The two most essential criteria for the redistribution problem are complying with the landowner preferences and redistributing the block areas evenly."

"While fixed facilities represent real estate like houses, gardens, etc, the landowner's preferences show the choices of each landowner that want to assign their parcels to the desired blocks."

"Once the process of GA is finished, the overflow and residual areas are completed with the help of an 8-step expert system to obtain the final land redistribution plan."

Implementation of DE and SS to Land Redistribution Problem

"While with the basic DE algorithm, a mutant individual is created, the gene values of two randomly selected individuals are subtracted and multiplied by the scaling factor F. This process will cause the continuous values to be converted into discrete values."

"Because the reference set is updated statically in this study, firstly, new individuals are obtained from the subsets with the combination method."

"In the uniform crossover, a random number is generated for each gene, and depending on whether this number is smaller or greater than 0.5, genes are selected from Individual 1 and Individual 2 to obtain the child."

"EO, MPA, and HBA methods run their search strategies as is for continuous search space, but their individuals are adapted to discrete search space by rounding process before evaluating with the fitness function."

Experimental Results and Comparisons

"When the SPP1 and SPP2 values are examined, it is seen that the expert system steps increased the success percentages of preference for the land redistribution plans obtained by the GA, HBA, and EO methods."

"The best redistribution plans based on fitness values (obtained for 2500 iterations over 30 runs for all GA, DE, SS, EO, MPA, and HBA methods) are used."

"DE, SS, and MPA, and the conventional methods got better than the GA, HBA, and EO techniques by the number of parcels."

"With the GA, SS, EO, and MPA techniques, while the maximum number of landowners with shares in a parcel is 10, this number is 11 with the DE and HBA methods and 13 with the conventional way."

"When the number of parcels per landowner was examined, it was seen that the SS, DE, and MPA methods were more successful when compared to the other methods."

Conclusion

"Nature-inspired algorithms, increasing in number each day, play an essential role in solving real-world problems."

"Due to the specific structures of real-world issues, and the differences in algorithms, more than one method was applied to the same problem, and their performances were analyzed."

"Six nature-inspired algorithms were used for a discrete real-world optimization problem."

"The other methods were not inadequate to compete with SS and DE methods, and they were seen to fail to overcome the complexity of the problem."

Matheuristics: Using Mathematics for Heuristic Design [2]

This is a machine-generated summary of:

Boschetti, Marco Antonio; Maniezzo, Vittorio: Matheuristics: using mathematics for heuristic design [2]

Published in: 4OR (2022)

Link to original: https://doi.org/10.1007/s10288-022-00510-8

If you want to cite the papers, please refer to the original.

For technical reasons we could not place the page where the original quote is coming from.

Abstract-Summary

"Matheuristics are heuristic algorithms based on mathematical tools such as the ones provided by mathematical programming, that are structurally general enough to be applied to different problems with little adaptations to their abstract structure."

"The result can be metaheuristic hybrids having components derived from the mathematical model of the problems of interest, but the mathematical techniques themselves can define general heuristic solution frameworks."

"We focus our attention on mathematical programming and its contributions to developing effective heuristics."

Introduction

"When a first international workshop was organized in Bertinoro, Italy, to discuss the use of mathematical tools for the design of heuristics [3], we needed a name to denote the specific focus of the event."

"We decided to use the term matheuristic because it is nice, it points very well to the focus on mathematics for developing heuristics, and Google returned at the time 0 found pages under that heading."

"In our recent book [4] we survey matheuristics from the viewpoint of metaheuristics, i.e., abstract, problem agnostic heuristic solving approaches."

"We introduce the main mathematical tools and then we describe how to use them for designing heuristics, reporting some contributions in the literature that provide interesting insights."

"The interested reader can find in the literature other surveys and special issues dedicated to Matheuristics (e.g., [5–8])."

Mathematical Programming

"In case, we can define an approximate model, accepting that the optimal solution found may not be optimal or feasible for the original problem, or we can consider the opportunity to heuristically solve the problem or to model only some subproblems within the heuristic algorithm."

"It is therefore computationally worthwhile to use these MIP models, instead of problem linear relaxations, to guide the MIP solvers in its most crucial steps, or anyway to solve to optimality MIP subproblems in order to speed up the solution of the whole problem we are interested in."

"Mentioning just a few contributions that closely stick to the idea of including MIP models in order to construct large-scale neighborhoods that are effectively explored by a black-box MIP solver to obtain high-quality solutions, we point out the works of De Franceschi et al. [9], Hewitt et al. [10] and Salari et al. [11]."

"KS is a matheuristic approach, which makes use of MILP solvers to obtain heuristic, possibly optimal, solutions of instances encoded as (mixed) integer linear programming problems."

Branch and Bound

"During this process, all nodes having a lower bound (for a minimization problem) greater or equal to the current best upper bound (i.e., the value of the best feasible solution found) can be eliminated, reducing the search space."

"A matheuristic approach specifically targeting branch and bound is Beam search (BS), whose central idea can be traced back to Lowerre [12]."

"BS does not complete the search that would normally be carried out by branch and bound algorithms, therefore it is an approximate method and a matheuristic of its own."

"At each BS iteration, the algorithm extends a partial solution from a set T, the beam, generating a possibly limited number of offsprings."

"A few major differences exist with respect to standard branch and bound; the main one being that search is not allowed to backtrack to partial solutions but it always restarts from scratch, thus no stack data structure is needed and search can be run in parallel."

Branch and Cut

"Both methods are self-sufficient, complete heuristics, but they came to be used as elective algorithms to be applied at some branch and bound nodes in order to improve the incumbent feasible partial solution."

"The selection is made by fixing all variables that have the same values in the bound and in the incumbent solutions, and letting the solver try to solve optimally the remaining MIP problem, that is called sub-MIP, within a given node limit or with an objective cutoff."

"The Corridor method (CM) is a general search method originally proposed by Sniedovich and Voß [13] as a dynamic programming (DP) heuristic overlay, and later extended beyond DP to other exact approaches, such as branch and bound."

"In its general form, CM tries to solve a possibly NP-hard optimization problem, for which we know an exact method (branch and bound, branch and cut, dynamic programming or else) that could effectively solve it on relatively small instances."

Dynamic Programming

"DP can be applied to more complex problems and in these cases each state can have a more complex structure (e.g., set, sequence, etc) and also the possible decisions can be more complex."

"In a DP for the Traveling Salesman Problem (TSP) the stage j could correspond to the length of the Hamiltonian path and the states are defined by the possible subsets of clients and the last client visited."

"DP algorithms can be used to compute bounds on the value of the optimal solution and in these cases some approximation can be also applied as, for example, the state-space relaxation [14]."

"Fore-and-back is a primal only method, but it is able to compute bounds to the cost of completing the partial solutions that are iteratively constructed, and it is therefore able to discard partial solutions from expansion thereby pruning the search trees."

Decomposition Methods

"Duality and decomposition methods provide some very useful data that can feed heuristic algorithms, in particular, dual variables and Lagrangian penalties have the power to gather information on the overall structure of the problem and of the specific instance to be solved."

"The authors propose a model for optimizing the implementation of the shrink operation as set partitioning problems with a side constraint, that is solved by a matheuristic that combines a dual ascent procedure, a Lagrangian pricing approach, and a Lagrangian heuristic."

"One such effort was presented in Boschetti and Maniezzo [15], where a 'bend-Heuristic' pseudocode suggested to use the standard Benders' decomposition and cutting planes insertion framework, applied to the LP relaxation of the problem of interest in case it is a combinatorial one, essentially suggesting to solve heuristically the master problem."

Metaheuristic Hybrids

"This section deals with metaheuristic algorithms that evolve one single solution, they are all essentially enhancements of a basic local search procedure."

"The algorithm can be very simple, as it only prescribes to start from a solution, find its local optimum with reference to a specific quality measure, perturb the incumbent solution, optimize this new one, and so on."

"Another timetable application is described in Gunawan et al. [16], where the authors design a simple matheuristic where an initial solution, obtained by solving a Lagrangian relaxation of the problem, is later improved by a SA."

"Gendron et al. [17], working on the multicommodity fixed-charge network design problem, propose to solve an LP relaxation of the problem and possibly fix it to get a heuristic solution."

"Yaghini et al. [18] use added cuts combining a cutting-plane neighborhood structure and a tabu search metaheuristic for the capacitated p-median problem."

Conclusions and Future Research Directions

"In this survey on Matheuristics we mainly consider the heuristic approaches that make use of linear programming, however the developments in nonlinear programming in recent years give new opportunities."

"Many mathematical tools for nonlinear programming are in fact not very effective for the exact solution of real-world problems, but they could be very useful for developing heuristic algorithms."

"An interesting observation is that there exist approaches described in the mathematical literature that have not yet proved effective enough to be used for the development of exact methods and, for this reason, they are not very popular today, but they could prove effective when designing heuristics."

"We also believe that the convergence between exact and matheuristic algorithms could be an interesting research direction, aimed at developing exact methods that can be easily transformed into heuristics and able to quickly generate feasible, near-optimal solutions."

Best Strategy to Win a Match: An Analytical Approach Using Hybrid Machine Learning-Clustering-Association Rule Framework [19]

This is a machine-generated summary of:

Srivastava, Praveen Ranjan; Eachempati, Prajwal; Kumar, Ajay; Jha, Ashish Kumar; Dhamotharan, Lalitha: Best strategy to win a match: an analytical approach using hybrid machine learning-clustering-association rule framework [19]

Published in: Annals of Operations Research (2022)

Link to original: https://doi.org/10.1007/s10479-022-04541-6

If you want to cite the papers, please refer to the original.

For technical reasons we could not place the page where the original quote is coming from.

Abstract-Summary

"One of the significant challenges in the sports industry is identifying the factors influencing match results and their respective weightage."

"For appropriate recommendations to the team management and the team players, there is a need to predict the match and quantify the important factors for which prediction models need to be developed."

"The second thing required is identifying talented and emerging players and performing an associative analysis of the important factors to the match-winning outcome."

"This paper formulates a hybrid machine learning-clustering-associative rules model."

"We predict the match outcome for One day Internationals (ODIs) and Twenty 20s (T20s) (two formats of Cricket representing 50 over and 20 over versions respectively) adopting state-of-the-art machine learning algorithms, Random Forest, Gradient Boosting, and Deep neural networks."

"The model is thus helpful for both team management and for players to improve their winning strategy and also for discovering emerging players to form an unbeatable team."

Introduction

"The impact of player-specific factors and environmental conditions like match time, match venue (home or away), batting position (whether the team bats first or second), and toss decisions are understated in current studies."

"Such data-backed identification would help understand how their performance impacts the winning outcome and identify associative rules that determine which combination of factors leads to a favorable match outcome."

"It is not driven by the captain's performance only but all 11 players and the impact of various categorical factors like the toss, whether the team is batting first or second, to impact match result."

"While current research primarily consists of statistical and operations research-based techniques, there is a proliferation of machine learning algorithms [20] that can predict the match result and quantify and identify the most critical factors influencing the results."

Literature Review

"The factor is antithetical to the number of runs scored since the fall of an important batsman can mold the outcome of the match accordingly in the direction of the bowling team [21]."

"This also depends on the pitch conditions and if a team is available to get a favorable toss outcome, this may influence the match outcome probabilistically and hence is included as a factor in the study [22]."

"There is a need to develop match result prediction models from more state-of-the-art machine learning techniques for higher accuracy and identify a weightage (importance score) to each factor influencing match outcome."

"Teams are driven by the performance of the captains and individual players, and the impact of various categorical factors like the toss, match time, and whether the team is batting first or second have an impact on match outcome."

Materials and Methods

"The predictors incorporated in the model are discussed below: The following appropriate factors, namely, Runs scored, wickets taken, Bowling Average, number of catches taken, Toss decision, Match venue/location, Match time, and Batting Position are considered for the predictive model."

"Step 2.2 The prediction accuracy and performance on 5% of the real data-points (150 data-points) from the ESPN Cricinfo website for the Indian players Hardik Pandya, Shreyas Iyer, Ravindra Jadeja, Rishabh Pant, and Virat Kohli are then compared to determine which of the three models (Random Forest, Gradient Boosting and Deep Neural networks) outperforms the others in match outcome prediction."

"Represent the 'm' number of inputs/predictors considered for the model representing the main (effect) variables Runs scored, wickets taken, Bowling Average, number of catches taken, Toss decision, Match venue/location, Match time, and Batting Position."

Analysis and Results

"Across all the machine learning models, the most significant factors impacting ODI match outcome are Player reputation and Match time (day and day/night), followed by the number of catches and runs scored."

"The above results are corroborated in the feature importance graph where Toss decision, 'Wickets*Batting position' and the triplet interaction variables 'Runs*Wickets*Catches' and 'Wickets*Catches*Match time' are the largest positive drivers of ODI match outcome."

"From Model 2, it is found that interaction variables 'Runs*Wickets', 'Runs*Match location', 'Wickets *Batting position', 'Catches *Match location' and 'Toss decision * Batting position' are significant implying and corroborating the result in Model 1."

"From the final dataset used for the modeling, variables such as runs, catches, bowling average, wickets, toss decision (win/loss), and match time (day/day and night) are considered."

Discussion

"Categorical environmental factors like toss result, match venue and batting order (whether the team is batting first or second) prove to be equally important as player specific statistics (quantitative factors) in determining match outcome."

"We found that the most significant factors impacting ODI match outcome as predicted by the DNN model are Player reputation and Match time (day and day/night), followed by the number of catches and runs scored."

"The vital factors for T20 match outcome are Batting order (first/second), toss decision, and runs scored are the most significant predictors, followed by player reputation."

"The match result prediction model enables the management to identify the various environmental factors and player parameters that influence the winning rate of a sports team."

"The match outcome prediction model enables the captains of the team to make appropriate decisions during toss and based on the condition of the pitch to maximize the winning rate."

Conclusion

"The paper has attempted to compare the predictive performance of random forest, gradient boost, and deep neural network-based models and present the significance of the factors in predicting the outcome of match instances scraped from ESPN CricInfo website."

"The deep neural network model is observed to outperform the other two machine learning models in terms of predictive accuracy and performance on unseen new players' data."

"The study is not without its limitations that can be worked upon in future research to generate additional insights in the area of sports outcome prediction."

"The data sample considered for model building is limited to 3000 instances."

"This sample size can be varied, and other parameters like the significance of partnerships, match-winning batting combinations can be considered predictors in the models."

A Nature Inspired Optimization Algorithm for VLSI Fixed-Outline Floorplanning [23]

This is a machine-generated summary of:

Shunmugathammal, M.; Columbus, C. Christopher; Anand, S.: A nature inspired optimization algorithm for VLSI fixed-outline floorplanning [23]

Published in: Analog Integrated Circuits and Signal Processing (2020)

Link to original: https://doi.org/10.1007/s10470-020-01598-w

If you want to cite the papers, please refer to the original.

For technical reasons we could not place the page where the original quote is coming from.

Abstract-Summary

"VLSI floorplan optimization problem aim to minimize the following measures such as, area, wirelength and dead space (unused space) between modules."

"This paper proposed a method for solving floorplan optimization problem using Genetic Algorithm which is named as 'Lion Optimization Algorithm' (LOA)."

"Although a number of GAs are developed for solving VLSI floorplan optimization problems, they are using weighted sum approach with single objective optimization and crossover between two B*tree structure is not yet attempted."

"This paper explains, power of B*tree crossover operator for multiobjective floorplanning problem."

Introduction

"Valenzuela and Wang [24] proposed GA for VLSI floorplanning problems."

"In this LOA paper, characters of Lion such as, mating, territorial defense and takeover, territorial marking and difference between life style of nomad and resident lions are used for solving VLSI floorplanning problems."

"Valenzuela and Wang [25] and Lin and others [26] presented genetic algorithms using polish expression to solve the VLSI floorplanning problem."

"Areibi [27] and Tang and Yao [28, 29] presented memetic algorithms for VLSI floorplan problems which has computationally expensive O-tree representation."

"After the detailed literature survey, crossover between two B*tree structures have not yet been proposed in GA for VLSI floorplan optimization problems."

"In this work, an attempt is also made to solve the modern non-slicing VLSI design floorplanning problem with the multiobjective of minimizing the silicon area and interconnecting wirelength along with the fixed outline constraint, using a genetic algorithm-based methodology resulting in minimum area and wirelength in considerable amount of time."

Lion's Social Behaviour

"Each pride consists of five or six females with their offsprings and one male or there can be up to two of them."

"Male lion pairs are more seen among nomads because who have been excluded from their maternal pride."

"Nomad lion may become resident and resident may become nomad [30, 31]."

"Both male and female lions in the same pride work as a team to protect each other and the young offsprings from other predators."

"There can be brutal battles between entire prides of lions over territory."

"Lion's social behavior begins when a new nomad male takes over a pride."

"In this takeover, the male lion kills all the existing offsprings so that it can reproduce right away."

"Mating and territorial takeover behavior of lions are used for optimization in VLSI floorplanning."

LOA for Floorplan Optimization

"They are, (1) Initial population Generation, which is responsible for generating different B*tree structures using mutation (2) Mating is enough to derive new B*tree structures (3) Territorial Defense and (4) Territorial Takeover is for finding new best B*tree fitness(cost) function and replacing worst fitness function by new best solution."

"Lion1 (initial B*tree) in T (initial population) are again partitioned into k subsets called prides." "Initial B*tree in ith pride is taken as lion (female) and 3rd B*tree in that same pride is taken as lion (male)."

"Crossover is done between initial B*tree (female lion) and 3rd B*tree (male lion)."

"Structural information between the initial (female lion) and 3rd B*tree (male) is interchanged by B*tree crossover."

"At the end of this territorial defense and territorial takeover, Lion Optimization Algorithm (LOA) identifies dominated B*tree (male lions) as optimal or acceptable solutions that play an important role in LOA."

Experimental Results and Discussions

"Performance of LOA is compared with following state-of-art algorithms, Thermal aware floorplan based Genetic Algorithm proposed by Hung and others [32], B*tree based Fast-SA proposed by Chen and Chang [33], Genetic Algorithm proposed by Singha and others [34], B*tree based Simulated spheroidizing annealing algorithm (SSAA) proposed by Anand and others [35], and Improved simulated spheroidizing annealing algorithm by Anand and others [35], B*tree-based Variable-Order Ant System (VOAS) proposed by Hoo and others [36], Particle Swarm Optimization–Genetic Algorithm (PSO-GA) based Hybrid algorithm proposed by Sivaranjani and Senthil Kumar [37] in terms of area and wirelength."

"Proposed LOA produces significant results in wirelength minimization and area minimization for MCNC bench marks."

"Proposed LOA is compared with other GA algorithms like Thermal aware floorplan-based GA proposed by Hung and others [32], Genetic algorithm proposed by Singha and others [34], PSO-GA Based Hybrid algorithm proposed by Sivaranjani and Senthil Kumar [37]."

"Proposed LOA produces best results for highest bench mark circuits in MCNC."

"For apte, xerox and hp LOA produces equal results compared to SA but the algorithms based on B*tree using SA doesn't have property of B*tree crossover."

"B*tree crossover used in proposed LOA produces a balance between global exploration and local exploitation as compared to simulated annealing algorithm."

"MCNC benchmark xerox produces best optimal solution compared to other well-known algorithms."

"The proposed LOA method, due to its crossover operator, it produces more converged pareto-solutions than SSAA."

Conclusion

"The proposed LOA is developed by the combination of B*tree crossover with genetic algorithm."

"LOA proved that this B*tree crossover is an efficient and effective technique for floorplan optimization using GA."

"LOA results are obtained using MCNC and GSRC benchmarks."

"Proposed LOA experimental results have been analysed using Analyse-it tool also."

"Statistical results show that proposed LOA produces significant results in wire-length minimization and area minimization for MCNC bench marks and also produces better results in dead space minimization for GSRC benchmarks while comparing the literature reports of well-known algorithms."

"A detailed study of inserting this B*tree crossover into other bio inspired algorithms as, Particle Swam Optimization (PSO), Ant Colony Optimization etc could be the future direction of this research."

Customized Simulated Annealing Based Decision Algorithms for Combinatorial Optimization in VLSI Floorplanning Problem [38]

This is a machine-generated summary of:

Anand, S.; Saravanasankar, S.; Subbaraj, P.: Customized simulated annealing based decision algorithms for combinatorial optimization in VLSI floorplanning problem [38]

Published in: Computational Optimization and Applications (2011)

Link to original: https://doi.org/10.1007/s10589-011-9442-y

If you want to cite the papers, please refer to the original.

For technical reasons we could not place the page where the original quote is coming from.

Abstract-Summary

"The aim of this work is to minimize the unused area, that is, dead space in the floorplan, in addition to these objectives."

"A Simulated Annealing Algorithm (SAA) based heuristic, namely Simulated Spheroidizing Annealing Algorithm (SSAA) has been developed and improvements in the proposed heuristic algorithm is also suggested to improve its performance."

"The performance of the proposed algorithm is compared with that of other stochastic algorithms reported in the literature and is found to be efficient in producing floorplans with very minimal dead space."

"The proposed SSAA algorithm is also found more efficient for problems of larger sizes."

Introduction

"The floorplanning problem has been addressed with the common objective of minimizing only the total chip area along with the interconnecting wirelength and all the classical floorplanning methodologies mostly handle only the block packing, that is, fixing the relative location of each module without any constraint on the floor outline."

"Paper optimization in floorplanning considering the fixed-die-outline constraint is attempted to minimize primarily the dead space in the floorplan layout in addition to minimizing the chip area and interconnecting wirelength."

"Chen and others [39] presented a B*tree based Simulated Annealing Algorithm, named as FastSA for designing fixed-outline floorplans, which resulted in a considerable amount of computational time."

"A SA-based algorithm is devised to find a near-to-optimal solution to the floorplanning problem using the fixed-outline constraint with the primary objective of minimizing the dead space."

Simulated Annealing Algorithm

"The probability depends on the difference of the solution quality and the temperature."

"In the classical annealing schedule, the temperature is reduced by a fixed ratio λ, for each iteration of the annealing process."

"Several annealing schemes have been proposed to reduce the running time of SAA."

"Chen and others [39] proposed FastSA algorithm, which reduces the running time of SAA compared to the older versions of SAA."

"The exploration done by FastSA algorithm may not be adequate enough to reach the near-to-optimal solution for a given VLSI floorplanning problem."

Proposed Simulated Spheroidizing Annealing Algorithm

"Like the metallurgical spheroidizing process, the proposed algorithm involves a local search process known as spheroidizing search for an extended period of time (that is, soaking time) at successive stages of the algorithm (that is at different temperatures)."

"During the search, whenever a neighborhood of a current solution is inferior, the difference in the objective function between the current and neighboring solution is computed as the uphill cost."

"If $\Delta cost_{avg}$ is smaller then, it implies that the cost of the neighboring solutions only changes a little; in that case, the temperature is still reduced to reduce the number of iterations."

"Even for a higher temperature, the SSAA was not able to produce better solution whenever a stage wherein the number of solutions rejected over the number of solutions searched at a temperature becomes a higher value of '0.99'."

Improved Simulated Spheroidizing Annealing Algorithm

"To further improve the performance of the proposed Simulated Spheroidizing Annealing Algorithm, the perturbation pattern, that is, the methods of finding the neighborhood solution during the spheroidizing search is modified."

"Swap two nodes/blocks) until it reaches the mth iteration of Improved Simulated Spheroidizing Annealing Algorithm (ISSAA) (in this work, m = 7)."

"The m and n values are selected from the trial and error simulations made until a satisfactory performance is obtained."

"The values m = 7 and n = 11 are used only for MCNC benchmarks."

"For GSRC benchmarks, values are set to m = 50 and n = 80."

Results and Discussion

"The proposed SSAA and ISSAA algorithms are tested with two kinds of benchmark circuits, namely, MCNC (Microelectronics Center of North Carolina) and GSRC (Gigascale Systems Research Center) to find solutions to modern floorplanning problems with the fixed-outline constraint."

"The performance of the proposed SSAA and ISSAA algorithms in terms of Area, Wirelength (WL), Dead Space (DS) and CPU time (CPUT) obtained on the MCNC and four GSRC benchmarks, were compared with that of other algorithms and heuristics such as SP(Parquet-1) [40], Parallel Simulated Annealing (PSA) algorithm [41], Timberwolf [42, 43], Classical SA [44], Discrete Particle Swarm Optimization (PSO) [45], Otree [46], Enhanced Otree [47], Transitive Closure Graph (TCG) [48], Corner Sequence (CS) [49], Genetic Algorithm [28], FastSP [50], Corner Block List (CBL) [51], and FastSA [39] reported recently in the literature."

"It is inferred that the proposed SSAA is as capable as the proposed ISSAA in finding nearer-to-optimal solutions for smaller benchmark circuits and ISSAA outperformes SSAA for larger circuits."

"It is Inferred that, the proposed algorithms SSAA and ISSAA are found equally efficient while solving larger size problems also."

"An attempt is also made to reduce the CPU time of the proposed methodologies through parallel computation."

Conclusion

"Modern VLSI floorplanning problems involve fixed-outline constraints and traditionally researchers have developed stochastic algorithms, which are mostly aimed at only minimizing the area occupied by the modules and the interconnecting wirelength between the modules."

"An SAA-based optimization algorithm, known as Simulated Spheroidizing Annealing Algorithm (SSAA), is proposed with the objective of minimizing primarily the dead space along with minimizing the total area and the wirelength between the modules."

"Of the experimentation, it is found that, though the proposed algorithms do not show significant improvement over other algorithms in terms of area and wirelength, but it surpasses all these algorithms in producing solutions with minimum dead space."

"The proposed algorithm is also improved by incorporating better methods of finding neighboring solutions during the search."

"An attempt is also made to minimize the CPU time of the proposed algorithm through parallel computing methodologies."

A PSO-Based Intelligent Decision Algorithm for VLSI Floorplanning [45]

This is a machine-generated summary of:

Chen, Guolong; Guo, Wenzhong; Chen, Yuzhong: A PSO-based intelligent decision algorithm for VLSI floorplanning [45]

Published in: Soft Computing (2009)

Link to original: https://doi.org/10.1007/s00500-009-0501-6

Abstract-Summary

“This paper proposes a novel intelligent decision algorithm based on the particle swarm optimization (PSO) technique to obtain a feasible floorplanning in VLSI circuit physical placement.”

“Inspired by the physics of genetic algorithm (GA), the principles of mutation and crossover operator in GA are incorporated into the proposed PSO algorithm to make this algorithm to break away from local optima and achieve a better diversity.”

“The proposed algorithm can avoid local minimum and performs well in convergence.”

“The experimental results of the proposed method in this paper can also greatly help floorplanning decision making in VLSI circuit design automation.”

Introduction

“Genetic algorithm (GA) has been proved to be an effective method for tackling NP-hard optimization problem [52], and has been successfully applied on VLSI floorplanning problems [25, 28, 53, 54].”

“The advantages of PSO over many other optimization algorithms are its simplicity in implementation and its ability to converge to a reasonably good solution quickly.”

“Literature [55] originally introduces PSO into the floorplanning problem.”

“The paper adopts the B*-tree floorplanning structure [56] to generate an initial stage with overlap free for placement and utilizes PSO to find out the potential optimal placement solution.”

"It is unable to solve problems that optimize the area and wire length simultaneously, thus it is not an effective PSO algorithm for general floorplanning problems."

"An intelligent decision algorithm based on discrete PSO (DPSO) for VLSI floorplanning is proposed, integer coding based on module number is also adopted in this method."

Problem Description

"The hard module's shape is fixed, and is denoted as (W, H), where W is the width and H is the height of the module."

"Soft module's area is also fixed, but the ratio of width/height is included in a given range."

"It can be denoted as (S, L, U), where S represents the area, L and U the lower and upper boundary of the width/height ratio."

"In case that the modules are given, the objective of VLSI floorplanning is to arrange the modules on a chip under the constraints that any two modules are not overlapped, and the area, wire length and other performance indices are optimal."

Particle Swarm Optimization and Its Parameter Analysis

"A suitable selection of the inertia weight balances the global and local exploration abilities and, consequently, reduces the number of iterations required to locate the optimum solution [57]."

"The search process of the PSO is nonlinear and very complicated, it is hard, if not impossible, to mathematically model the search process to dynamically adjust the inertia weight [58]."

"Using a relatively large value to a small value through the course of a PSO run, PSO tends to have more global search ability at the beginning of the run while having more local search ability near the end of the run [57, 59]."

"A fuzzy system was designed to adjust the inertia weight of PSO dynamically [58]."

"These improved PSO all discussed the inertia weight as a global variable."

PSO-Based Intelligent Decision Algorithm for Floorplanning

"Considering the minimization of area, the fitness is equal to the value of area, thus the fitness function can be represented as follows. Where W(x) is the floorplan width of the corresponding particle x, and H(x) is the floorplan height of x. The code x stands for the module sequence, and some rules are needed to convert modules sequence to two-dimensional floorplan."

"Considering the constraint mentioned above, module j should be placed at point b. When it comes to optimize the area and total wire length simultaneously, the floorplanning performs a multi-objective minimization."

"For this layout, if the problem contains m nets, its total wire length is The details of algorithm are described as follows: Step 1: Load modules data and initial the parameters of the PSO algorithm (such as population size, generations, etc)."

Experimental Results

"$C_1 = 0.82$–0.5, $C_2 = 0.4$–0.83 are considered as the optimal parameter setting modal and applied in the algorithm of this paper."

"In order to show the validity of the algorithm, we have compared the results with Chang et al. [56] and Sun et al. [55]."

"The experiment results show that the proposed algorithm can gain better result than Chang et al. [56]."

"In case that the problem scale is small, the proposed method has a similar performance compared with literature [55]."

"In case that the problem scale becomes larger, our algorithm has a better performance."

"For xerox, ami33 and ami49 benchmarks, while the areas obtained are similar to those in compared algorithms, the wire lengths are little worse."

Conclusion

"The experimental results demonstrated that the proposed algorithm can achieve the optimal and reasonable solution for the hard IP modules placement and also can be applied to multi-objective problem in floorplanning."

"For single-objective optimization, our algorithm performed well."

"For multi-objective optimization, the results obtained were better in some benchmarks and not well in other benchmarks."

"These objectives may be hard to weight because of different properties."

"To avoid these weaknesses, our future work will focus on enhancing the efficiency of the proposed algorithm and finding ways to apply pareto optimality to the multi-objective problem in floorplanning."

Thermal-Aware Non-slicing VLSI Floorplanning Using a Smart Decision-Making PSO-GA Based Hybrid Algorithm [37]

This is a machine-generated summary of:

Sivaranjani, P.; Senthil Kumar, A.: Thermal-Aware Non-slicing VLSI Floorplanning Using a Smart Decision-Making PSO-GA Based Hybrid Algorithm [37]

Published in: Circuits, Systems, and Signal Processing (2015)

Link to original: https://doi.org/10.1007/s00034-015-0020-x

If you want to cite the papers, please refer to the original.

For technical reasons we could not place the page where the original quote is coming from.

Abstract-Summary

"A smart decision-making hybrid particle swarm optimization-genetic algorithm that aims at reducing the area, wirelength, and hotspot by distributing the temperature evenly across the chip is presented."

"Temperature-driven floorplanning is considered at the perturbation stage to separate the hotspots, thereby reducing the average and maximum temperature."

"The experimental results of the proposed algorithm are compared with other stochastic algorithms using MCNC and Alpha processor floorplan benchmark circuits."

"The result shows that the proposed algorithm performs efficient floorplanning, with reduced average and peak temperature."

Introduction

"According to Moore's law [60], the number of transistors doubles every 18 months, which increases the complexity of the circuits, and at the same time, the feature size of the IC is also considerably scaled down."

"Scaling of transistors cause the CMOS feature size to shrink to nanometer scale, leading to an increase in power density."

"As the power density increases, there will be an increase in heat dissipation (hotspot) and finally an increase in temperature."

"The increase in temperature is mainly due to small feature size, higher packing density, and increase in power consumption."

"Due to high integration, the rate of power consumption in different modules of floorplan (e.g., the power density) varies significantly."

"The peak temperature and large thermal gradient also affects the different aspects of interconnect design including signal delay, routing, congestion and reliability [61]."

Related Work

"O-tree representation is proposed [62] based on a non-slicing floorplan."

"The iterative method incorporates a metaheuristic approach, such as simulated annealing, a genetic algorithm or a particle swarm optimization algorithm to find the final optimal floorplan."

"A novel PSO-GA based hybrid algorithm [63] is more effective for the multimodal optimization of NP-hard problems."

"Since VLSI floorplanning is a NP-hard problem with multi-objective function, a PSO-GA based hybrid algorithm is adopted in this work."

"In [64], the authors scheduled a genetic algorithm to tackle the VLSI floorplanning problem."

"A new heuristic method [65] was proposed that applies hybrid simulated annealing (HSA) to represent a non-slicing floorplan with an objective function by restricting the area and wirelength."

"The authors of [66, 67] proposed genetic algorithm(GA)-based thermal-aware floorplanning."

"Geometric programming (GP) [68] has been proposed for the optimization problem of temperature reduction and chip area floorplanning only."

Problem Statement

"The floorplan can be represented as a slicing or a non-slicing floorplan."

"Slicing floorplan can solve the floorplanning problems with limited scope such as increase in layout area and deadspace."

"The non-slicing floorplan is more general and can represent any type of packing with minimal layout area and deadspace."

"Due to this reason, many researchers are focused on the non-slicing floorplan representation."

B-Tree Representation*

"It is an ordered binary tree where the root node corresponds to the module on the bottom-left."

"The construction of B*-tree is similar to the depth-first search (DFS) algorithm."

"In B*-tree, starting from the root node, construct the left subtree and then the right subtree in a recursive fashion, such that the root node represents the module at the bottom-left (BL) corner in the floorplan."

Hybrid Particle Swarm Optimization: Genetic Algorithm

"The objective of the genetic algorithm is to find an optimal solution."

"In the modified genetic algorithm, each chromosome is given a dying probability based on the living generation."

"In PSO-GA based hybrid algorithm, each particle and chromosome corresponds to a potential solution."

"The algorithm starts with a randomly generated initial population where each particle generates a floorplan."

"The position of each block is obtained from the algorithm, which is given as an input file to Hotspot, whose other input is the power trace file."

"The fitness is assigned to each particle based on the parameter area, wirelength and temperature."

"Algorithm description: Step 1: Generate an initial population using B*-tree representation, load the module input and initialize the parameter of the PSO-GA based hybrid algorithm."

"The stopping condition for the proposed algorithm is as follows: (i) If fitness value stagnates for designated number of iterations. (ii) Fixed number of generations is reached."

Simulation Results

"The peak temperature of a block in a chip depends not only on its own power density, but also on the power density of the neighboring adjacent blocks."

"The placement of building blocks or a floorplan selected for a given chip will affect the maximum temperature of the chip considerably."

"In hotspot, usually the blocks with high power density have the highest temperature, but this is not always true."

"The temperature of a block in a chip depends not only on its power density but also on the power density of the adjacent block."

"For the same total value of power dissipation and the same floorplan layout, different distribution of power among various blocks can lead to altogether different temperature profiles of the chip."

"The optimal peak temperatures for the different benchmark circuits are obtained with minimal area."

Conclusion and Future Work

"The performance analysis using Alpha processor and MCNC benchmark circuits establishes that the temperature optimization technique decreases the chip temperature for all benchmark circuits while providing a compact floorplan compared to other traditional area-oriented techniques."

"The efficiency of the floorplan can be enhanced by including crossover operation in B*-tree representation; this can be considered for further research."

"For handling multi-objective optimization, weighted sum method is used to convert into single objective."

"Pareto optimality can be considered for dealing with multi-objective optimization."

References

1. Hakli H, Uguz H, Ortacay Z (2022) Comparing the performances of six nature-inspired algorithms on a real-world discrete optimization problem. Soft Comput 26:11645–11667. https://doi.org/10.1007/s00500-022-07466-1
2. Boschetti MA, Maniezzo V (2022) Matheuristics: using mathematics for heuristic design. 4OR-Q J Oper Res 20:173–208. https://doi.org/10.1007/s10288-022-00510-8
3. Maniezzo V (2006) Matheuristics 2006 conference web portal. http://astarte.csr.unibo.it/Matheuristics2006/
4. Maniezzo V, Boschetti M, Stuezle T (2021) Matheuristics: algorithms and implementations, EURO advanced tutorials on operational research. Springer, Berlin
5. Boschetti MA, Maniezzo V, Roffilli M, Bolufé Röhler A (2009b) Matheuristics: optimization, simulation and control. In: Blesa M, Blum C, Di Gaspero L, Roli A, Sampels M, Schaerf A (eds) Hybrid metaheuristics, HM 2009. Lecture notes in computer science, vol 5818. Springer, Berlin
6. Fischetti M, Fischetti M (2018) Matheuristics. In: Marti R, Pardalos PM, Resende MGC (eds) Handbook of heuristics. Springer, Cham. https://doi.org/10.1007/978-3-319-07124-4_14
7. Maniezzo V, Stützle T (2020) Special issue: matheuristics and metaheuristics. Int Trans Oper Res 27:1
8. Maniezzo V, Stützle T, Voß S (2009) Matheuristics: hybridizing metaheuristics and mathematical programming, Annals of information systems, vol 10. Springer, Berlin
9. De Franceschi R, Fischetti M, Toth P (2006) A new ILP-based refinement heuristic for vehicle routing problems. Math Program B 105(2–3):471–499
10. Hewitt M, Nemhauser GL, Savelsbergh MWP (2010) Combining exact and heuristic approaches for the capacitated fixed-charge network flow problem. Informs J Comput 22(2):314–325

11. Salari M, Toth P, Tramontani A (2010) An ILP improvement procedure for the open vehicle routing problem. Comput Oper Res 37(12):2106–2120
12. Lowerre B (1976) The HARPY speech recognition system. PhD thesis, Carnegie Mellon University, Pittsburgh, PA
13. Sniedovich M, Voß S (2006) The corridor method. A dynamic programming inspired metaheuristic. Control Cybern 35(3):551–578
14. Christofides N, Mingozzi A, Toth P (1981) State-space relaxation procedures for the computation of bounds to routing problems. Networks 11(2):145–164
15. Boschetti MA, Maniezzo V (2009b) Benders decomposition, Lagrangian relaxation and metaheuristic design. J Heuristics 15(3):283–312
16. Gunawan A, Ming Ng K, Leng Poh K (2012) A hybridized Lagrangian relaxation and simulated annealing method for the course timetabling problem. Comput Oper Res 39(12):3074–3088
17. Gendron B, Hanafi S, Todosijević R (2016) An efficient matheuristic for the multicommodity fixed-charge network design problem. IFAC PapersOnLine 49(12):117–120
18. Yaghini M, Karimi M, Rahbar MA (2013) Hybrid metaheuristic approach for the capacitated p-median problem. Appl Soft Comput 13(9):3922–3930
19. Srivastava PR, Eachempati P, Kumar A, Jha AK, Dhamotharan L (2022) Best strategy to win a match: an analytical approach using hybrid machine learning-clustering-association rule framework. Ann Oper Res 325(1):1–43. https://doi.org/10.1007/s10479-022-04541-6
20. Deval G, Hamid F, Goel M (2021) When to declare the third innings of a test cricket match? Ann Oper Res 303:81–99
21. Thorley J (2021) Age-related changes in the performance of bowlers in Test match cricket. Int J Sports Sci Coaching 16(5):1138–1151
22. Sahu A (2021) Predictive analysis of cricket. Turk J Comput Math Educ 12(6):5111–5124
23. Shunmugathammal M, Columbus CC, Anand S (2020) A nature inspired optimization algorithm for VLSI fixed-outline floorplanning. Analog Integr Circ Sig Process 103:173–186. https://doi.org/10.1007/s10470-020-01598-w
24. Christine Valenzuela L, Wang PY (2000) A genetic algorithm for VLSI floorplanning. In: Proceedings of the 6th international conference on parallel problem solving from nature, Paris, France. IEEE, pp 671–680
25. Christine Valenzuela L, Wang PY (2002) VLSI placement and area optimization using a genetic algorithm to breed normalized postfix expressions. IEEE Trans Evol Comput 6(4):390–401
26. Lin T, Chen DS, Wang YW (2004) Robust fixed-outline floorplanning through evolutionary search. In: ASP-DAC'04: proceedings of the 2004 Asia and South Pacific design automation conference. Piscataway, IEEE Press, pp 42–44
27. Areibi S (2001) Recursive and flat partitioning for VLSI circuit design. In: Proceedings of the 13th international conference on microelectronics, Rabat, Morocco. IEEE, pp 237–240
28. Tang M, Sebastian M (2005) A genetic algorithm for VLSI floorplanning using O-tree representation. In: Applications on evolutionary computing, Lecture notes in computer science, 3449. Springer, Berlin, pp 215–224
29. Tang M, Yao X (2007) A memetic algorithm for VLSI floorplanning. IEEE Trans Syst Man Cybernet 37(1):123–135
30. Maziar Y, Fariborz J (2016) Lion optimization algorithm (LOA): a nature-inspired metaheuristic algorithm. J Comput Des Eng 3:24–36
31. Rajakumar BR (2012) The Lion's Algorithm: a new nature-inspired search algorithm. In: Proceedings of 2nd international conference on communication, computing & security (ICCCS-2012), vol 6. Procedia Technology, pp 126–135
32. Hung WL, Xie Y, Vijaykrishnan N, Addo-Quaye C, Theocharides T, Irwin MJ (2005) Thermal-aware floorplanning using genetic algorithms. In: Sixth international symposium on quality electronic design (ISQED'05). IEEE. https://doi.org/10.1109/isqed.2005.122
33. Chen TC, Chang YW (2006) Modern floorplanning based on fast simulated annealing. IEEE Trans CAD 25(4):510–522

34. Singha T, Dutta HS, De M (2012) Optimization of floorplanning using genetic algorithm. Proc Technol 4:825–829
35. Anand S, Saravanasankar S, Subbaraj P (2012) Customized simulated annealing-based decision algorithms for combinatorial optimization in VLSI floorplanning problem. Comput Optim Appl 52(3):667–678
36. Hoo CS, Jeevan K, Ganapathy V, Ramaiah H (2013) Variable-order ant system for VLSI multiobjective floorplanning. Appl Soft Comput 13(7):3285–3297
37. Sivaranjani P, Senthil Kumar A (2012) Thermal-aware non-slicing VLSI floorplanning using a smart decision-making PSO-GA based hybrid algorithm. Circuits Syst Signal Process 34:3521–3542. https://doi.org/10.1007/s00034-015-0020-x
38. Anand S, Saravanasankar S, Subbaraj P (2011) Customized simulated annealing based decision algorithms for combinatorial optimization in VLSI floorplanning problem. Comput Optim Appl 52(3):667–689. https://doi.org/10.1007/s10589-011-9442-y
39. Chen TC, Chang YW (2006) Modern floorplanning based on B*-tree and fast simulated annealing. IEEE Trans CAD 25(4):510–522
40. Adya SN, Markov IL (2003) Fixed—outline floorplanning: enabling hierarchical design. IEEE Trans Very Large Scale Integr Syst 11(6):1120–1135
41. Fang JP, Chang YL, Chen CC, Liang WY, Hsieh TJ, Satria M, Han CC (2009) A parallel simulated annealing approach for floorplanning in VLSI. In: Algorithms and architectures for parallel processing, Lecture notes in computer science, vol 5574. Springer, Berlin, pp 291–302
42. Sechen C, Sangiovanni-Vincentelli A (1985) The timberwolf placement and routing package. IEEE J Solid-State Circuits 20(2):510–522
43. Sechen C, Sangiovanni-Vincentelli A (1986) Timberwolf3.2: a new standard cell placement and global routing package. In: DAC '86: Proceedings of the 23rd ACM/IEEE Design Automation Conference. IEEE Press, Piscataway, pp 432–439
44. Kirpatrick S, Gelatt C, Vecchi M (1983) Optimization by simulated annealing. Science 220(4598):671–680
45. Chen G, Guo W, Chen YA (2010) A PSO-based intelligent decision algorithm for VLSI floorplanning. Soft Comput Fusion Foundation Method Appl 14(12):1329–1337. https://doi.org/10.1007/s00500-009-0501-6
46. Guo PN, Cheng CK, Yoshimura T (1999) An O-tree representation of non-slicing floorplan and its applications. In: DAC '99: Proceedings of the 36th annual ACM/IEEE Design Automation Conference. ACM, New York, pp 268–273. https://doi.org/10.1145/309847.309928
47. Pang Y, Cheng CK, Yoshimura T (2000) An enhanced perturbing algorithm for floorplan design using the o-tree representation. In: ISPD '00: Proceedings of the 2000 International Symposium on Physical Design. ACM, New York, pp 168–173. https://doi.org/10.1145/332357.332395
48. Lin JM, Chang YW (2001) TCG: a transitive closure graph-based representation for non-slicing floorplans. In: Proceedings of the 38th Conference on Design Automation, Las Vegas, Nevada. ACM, New York, pp 764–769
49. Lin JM, Chang YW, Lin SP (2003) Corner sequence-A P-admissible floorplan representation with a worst case linear-time packing scheme. IEEE Trans Very Large Scale Integr Syst 11(4):679–686
50. Tang X, Wong DF (2001) FAST-SP: a fast algorithm for block placement based on sequence pair. In: Proceedings of the 2001 Asia and South Pacific Design Automation Conference, ASP-DAC '01. ACM, New York, pp 521–526. https://doi.org/10.1145/370155.370523
51. Hong X, Huang G, Cai Y, Gu J, Dong S, Cheng CK, Gu J (2000) Corner block list: an effective and efficient topological representation of non-slicing floorplan. In: Proceedings of the 2000 IEEE/ACM International Conference on Computer-Aided Design, ICCAD '00. IEEE Press, Piscataway, pp 8–12. http://portal.acm.org/citation.cfm?id=602902.602905
52. Goldberg DE (1989) Genetic algorithm in search optimization, and machine learning. Addison-Wesley, Reading

53. Gwee B, Lin M (1999) A GA with heuristic based decode for IC floorplanning. Integr VLSI J 28(2):157–172
54. Tang M, Yao X (2007) A memetic algorithm for VLSI floorplanning. IEEE Trans Syst Man Cybern 37(1):62–69
55. Sun TY, Hsieh ST, Wang HM, Lin CW (2006) Floorplanning based on particle swarm optimization. In: 2006 IEEE computer society annual symposium on VLSI. IEEE, Karlsruhe, pp 7–11
56. Chang YC, Chang YW, Wu GM, Wu SW (2000) B*-tree: a new representation for non-slicing floorplans. In: Proceedings of the 37th conference on design automation. ACM, Los Angeles, CA, pp 458–463
57. Shi YH, Eberhart RC (1998) Parameter selection in particle swarm optimization. In: Eiben A, Porto V, Saravanan N, Waagen D (eds) Evolutionary programming VII. Springer, San Diego, pp 591–600
58. Shi YH, Eberhart RC (2001) Fuzzy adaptive particle swarm optimization. In: Proceedings of the IEEE congress on evolutionary computation, Seoul. IEEE Press, New York, pp 101–106
59. Shi YH, Eberhart RC (2000) Experimental study of particle swarm optimization. In: Proceedings of the world multiconference on systemics, cybernetics and informatics 2000, Orlando, FL. IEEE
60. Moore G-E (1998) Cramming more components onto integrated circuits. In: Proceedings of the IEEE. IEEE, pp 82–85
61. Winther AT, Liu W, Nannarelli A, Vrudhula S (2011) Temperature dependent wire delay estimation in floorplanning. In: NORCHIP, 2011. IEEE, pp 1–4
62. Guo P-N, Cheng C-K, Yoshimura T (1999) An O-tree representation of non-slicing floorplan and its applications. In: Proceedings of the 36th Annual ACM/IEEE Design Automation Conference. ACM, pp 268–273
63. Shi X, Liang Y, Lee H, Lu C, Wang L (2005) An improved GA and a novel PSO-GA-based hybrid algorithm. Inf Process Lett 93(5):255–261
64. Tang M, Sebastian A (2005) A genetic algorithm for VLSI floorplanning using O-tree representation. In: Rothlauf F, Branke J, Cagnoni S, Corne DW, Drechsler R, Jin Y, Machado P, Marchiori E, Romero J, Smith GD, Squillero G (eds) Applications of evolutionary computing. Springer, Berlin, pp 215–224
65. Chen J, Zhu W, Ali M (2011) A hybrid simulated annealing algorithm for nonslicing VLSI floorplanning. IEEE Trans Syst Man Cybern C Appl Rev 41(4):544–553
66. Gracia N, Rajaram S, Nivethitha N, Sudarsan A (2012) Thermal aware modern VLSI floorplanning. In: 2012 International Conference on Devices, Circuits and Systems (ICDCS). (IEEE, pp 187–190
67. Hung W-L, Xie Y, Vijaykrishnan N, Addo-Quaye C, Theocharides T, Irwin MJ (2005) Thermal-aware floorplanning using genetic algorithms. In: Sixth International Symposium on Quality of Electronic Design, 2005 (ISQED 2005). IEEE, pp 634–639
68. Chen Y-C, Li Y (2010) Temperature-aware floorplanning via geometric programming. Math Comput Model 51(7):927–934

Chapter 2
High-Level Synthesis in VLSI Design

Introduction by the Editor

High-level Synthesis is a step used in VLSI design to convert an architecture behavior model, typically coded in C, C++ or SystemC, to a Register transfer language (RTL) model. This step is profoundly used in front end cycle of chip design in VLSI industry. RTL model is essentially a Verilog or VHDL model. Industry uses Synopsys Fusion Compiler (FC) or Synopsys Design Compiler (DC) for HLS. HLS helps implement the architecture's function in a way which consumes least power, least area and maximizes performance. This chapter circles various HLS methods used in VLSI. The techniques discussed estimate the operation age of an integrated circuit, solve Boolean Satisfiability for computer-aided design problem and summarize methods in integrated circuit reverse engineering with Latency-sensitive HLS for complex DSP Design. Gate sizing with pipelining constraints and geometric programing is also discussed in this chapter. Crosstalk noise mitigation through variable gate bias transmission gate is covered.

Remaining useful lifetime (RUL) is inversely proportional to failure probability of an IC. Operational is largely derived from RUL. Also, IDDQ testing, also known as quiescent state testing, can also help determine operational age of an IC. IDDQ current is the current drawn by the IC in its quiescent state or steady state. This decreases with extensive IC usage. The current decrease is due formation of interface traps and positive oxide layer formation due to negative gate bias voltage over time. This typically happens in PMOS. This is called Negative Bias Temperature instability (NBTI). In an IC, if we measure IDDQ for a large set of always on transistors and a large set of sparsely on transistors and check delta between it over a period, it is observed that the difference increases due to IC wear and tear.

Once architecture goes through HLS, it's critical to check for logic equivalence. Boolean Satisfiability (SAT) can prove efficient. Solving for satisfiability is finding such values a Boolean expression that can make the expression true. Combining this

A. S. Shastri (ed.), *Optimization Methods in VLSI Design*,
https://doi.org/10.1007/978-981-95-2457-0_2

"This current decreases as the circuit ages, largely due to the increase in the magnitude of the PMOS transistor threshold voltage caused by negative bias temperature instability (NBTI)."

"Novelty of our technique is a normalized difference, ΔI, computed from current measurements at two input test patterns and is proposed as a self referencing circuit age indicator."

"The difference between these two I_{DDQ} values increases with the circuit age."

"This approach requires no hardware modification in the circuit and, hence, can be applied to legacy ICs."

Introduction

"We propose a novel method of detecting aged ICs by measuring the quiescent leakage current of the circuit, referred to as I_{DDQ}."

"We exploit the change in transistor threshold voltages caused by Negative Bias Temperature Instability (NBTI) [2, 3] due to the operational stress during the chip lifetime in powered up state."

"Unused chips are expected to display only minimal threshold voltage changes since manufacture, while the PMOS transistors in the used parts will display varying increases in threshold voltage depending on the level of operational stress experienced."

"We use the externally measured I_{DDQ} for the entire chip to track aggregate shifts in threshold voltages for large number of transistors since it is impractical to directly measure device parameters inside an IC."

"I_{DDQ} decreases with age because the transistor threshold voltages increase resulting in reduced leakage from OFF transistors."

Prior Work

"Zhang and others [4] proposed a fingerprint based on path-delays in a chip to detect recycled ICs."

"Zheng and others [5] proposed a characterization method based on extraction of scan path delay signatures for a chip."

"They isolated the unstable and most age sensitive cells to devise a recycled IC detection method."

"On-chip ring oscillators (RO) are considered by several researchers to detect recycled ICs."

"Saneyoshi and others [6] proposed a hybrid on-chip age monitor containing RO and delay line."

"Zhang and others [7, 8] proposed a lightweight on-chip sensor using ring oscillators (ROs) to detect recycled ICs."

"He and others [9] proposed another lightweight on-chip design to exploit electromigration-induced aging effect of interconnect wires."

"Guin and others [10] have proposed an approach that uses RO and a digital signature to protect the RO frequency from tampering such that a recycled IC is accurately identified."

[Section 3]

"The I_{DDQ} drawn from the power supply is made up of the sub-threshold leakage currents controlled by OFF transistors and the gate oxide leakage currents in the transistors that are ON."

"That as the threshold voltage of a PMOS/NMOS transistor increases due to aging, the leakage current I_{DDQ}, which has a negative exponential relationship with the threshold voltage (V_{th}), decreases [11]."

"In a regular operation, the node with a higher probability of 0 (low 1-controllability) receives 0 more frequently and ages the next gate faster compared to a gate with an input of high 1-controllability."

"The patterns T_1 and T_2 are so selected that I_1 is largely controlled by PMOS devices of fast aging gates and I_2 by PMOS devices of slow aging gates."

"Pattern T_1 applies 11 to fast aging gates so that their PMOS transistors will control the leakage."

A Method for Detecting Recycled ICs

"To characterize, we will derive two test patterns for I_{DDQ} measurement and a threshold value for ΔI_{DDQ}."

"During the test, we measure the I_{DDQ} for the two selected test patterns, and a decision is made based on the normalized ΔI_{DDQ} value."

"This is done by the chip manufacturer through pattern selection and threshold calculation."

"The second part of the characterization process calculates a threshold value to determine whether or not a chip is recycled."

"Apply input patterns T_1 and T_2 to all four netlists and measure I_{DDQ}."

"The maximum value of ΔI found in Step 3 is selected as the threshold value ΔI_T for detecting recycled chips."

"In the foundry, two previously selected input patterns, T_1 and T_2, can be applied to a reasonably large number of ICs and ΔI measured."

"Input patterns T_1 and T_2 are applied to the chip under test."

Results and Discussion

"We performed aging simulation for several ISCAS'85 benchmark circuits [12]."

"The aging simulation assumed 25 °C temperature and a nominal 1 V supply voltage."

"Maximum value of ΔI, which is the threshold (ΔI_T) for each circuit is shown in Column 15."

"For c432 benchmark circuit, ΔI values in new circuit for four netlists representing process corners, are 3.30%, 3.31%, 3.15% and 3.45%, respectively."

"Column 15 here gives the minimum values of ΔI obtained from the four netlists."

"For c432, after 6 months of aging, ΔI values are 6.07%, 6.09%, 5.89% and 6.37%, respectively."

"That ΔI will further increase when the circuit is aged beyond 1 year."

"The other effect of temperature is related to the rate of aging degradation in the device itself when it is used at temperatures different from the nominal 25 °C."

Conclusion

"In one of the two test patterns I_{DDQ} is controlled by the least aged transistors, while in the other pattern it is controlled by the most aged transistors."

"For non-scan synchronous circuits and asynchronous circuits, the generation and application of a suitable pair of patterns are not as straightforward and require research."

"In the area of testing, very large circuits present a problem for I_{DDQ} based methods."

"How well the I_{DDQ} based recycled IC detection will work for large circuits should be investigated."

"Besides analyzing large circuits, our plans include actual hardware tests using the available Advantest T2000GS ATE at Auburn University."

"It is reasonable for the future to assume that the two-pattern aging test could become a part of the device specification."

"For the old legacy devices, some perhaps no longer in production, one must face the challenge to dig out from a design house or foundry the information needed to generate the aging tests."

Crosstalk Noise Mitigation Using a Transmission Gate with Varied Gate Bias [13]

This is a machine-generated summary of:

Sayil, Selahattin; Lamichhane, Subed; Sayil, Kutay: Crosstalk noise mitigation using a transmission gate with varied gate bias [13]

Published in: Analog Integrated Circuits and Signal Processing (2020)
Link to original: https://doi.org/10.1007/s10470-020-01706-w

If you want to cite the papers, please refer to the original.

For technical reasons we could not place the page where the original quote is coming from.

Abstract-Summary

"Driver sizing is a practical and a feasible technique for controlling the crosstalk noise in the post route design stage."

"Although an increased victim driver size can easily mitigate the coupled noise, it may create additional crosstalk issues with neighboring wires."

"Sizing up a gate results in an increased circuit area and affects the timing of paths through the gate."

"This work proposes a new crosstalk noise mitigation methodology based on our novel transmission gate design."

Introduction

"Driving or gate sizing method is a practical crosstalk noise reduction technique in the post route design stage [14]."

"We propose a new crosstalk noise mitigation methodology based on a novel transmission gate (TG) design with minimum area and performance overheads."

"Speaking, SET filtering methods such as driver sizing (on-site filtering) and low pass filtering techniques (along the data path) [15, 16] could also be applied towards interconnect crosstalk mitigation."

"The proposed crosstalk mitigation method, however, is based on applying different gate biases on the pass transistors of a TG depending on the crosstalk magnitude that needs to be mitigated."

"For faster designs, the technique can be combined with the driver sizing method in mitigating the crosstalk pulse to keep delay overhead more controlled."

"Gate sizing is used partially suppress the propagated noise at the output of the victim driver while the TG suppresses the remaining crosstalk glitch."

Transmission Gate Filter Design

"The proposed method in this work is based on applying different gate biases on the TG pass transistors depending on the crosstalk noise magnitude that needs to be mitigated."

"For the basic TG method, the gate voltages are taken as $V_{GN} = 1.2$ V (V_{DD}), and $V_{GP} = 0$ V. For the proposed method, the gate voltages of NMOS and PMOS transistors are symmetrically varied from their respective values using $V_{GN} = V_{DD} - \Delta V$ and $V_{GP} = \Delta V$, where $\Delta V > 0$."

"The strong filtering effect can be attributed to increased resistance of the TG due to reduced gate-source (source-gate) voltages of NMOS (PMOS) pass transistors."

"The sizes of M1 and M2 transistors need be adjusted to give desired values of gate voltages."

"The TG method requires two additional supply voltages and associated routing resources."

The Proposed Crosstalk Mitigation Method: TG Based Mitigation Combined with Driver Sizing

"The TG technique can be combined with driver sizing method to keep delay penalty low and the crosstalk mitigation is shared between the TG and the driver sizing."

"Victim driver sizing method is used partially to suppress the crosstalk noise while the TG suppresses the remaining pulse."

"Our method relies on incrementing the victim driver size by a certain amount (in our case by 1×) and then choosing the proper gate voltages of TG to eliminate the remaining transient."

"The analysis results show that the basic TG method always has more area overhead compared to driver sizing taken alone in most cases."

"For all 500 cases considered, the proposed approach results in on average a 41% area savings compared to victim driver sizing method taken alone."

"For the combined method, the simulation results have indicated that a 1× increase on victim driver size was sufficient to keep delay at manageable at our given slack."

Conclusion

"This work proposed a TG based technique for mitigating crosstalk noise."

"This approach results in smaller area overhead compared to driver sizing taken alone."

"The delay overhead of our novel TG is proportional to crosstalk noise width."

"This work proposed a combined approach where driver sizing is combined with the TG in order to keep delay more controlled."

"The proposed method resulted in about 41% area savings on average for all cases considered compared to driver sizing technique."

Boolean Satisfiability Methods for Modern Computer-Aided Design Problems in Microelectronics [17]

This is a machine-generated summary of:

Zapletina, M. A.; Zhukov, D. V.; Gavrilov, S. V.: Boolean Satisfiability Methods for Modern Computer-Aided Design Problems in Microelectronics [17]

Published in: Russian Microelectronics (2021)

Link to original: https://doi.org/10.1134/s1063739721070131

If you want to cite the papers, please refer to the original.

For technical reasons we could not place the page where the original quote is coming from.

Abstract-Summary

"The Boolean satisfiability (SAT) methods are one of the efficient approaches used to solve the problems of Boolean matching and the equivalence checking of digital circuits."

"In combination with classic routing algorithms and optimization techniques, these methods demonstrate better results than the classic routing algorithms in terms of the speed of operation and the quality of the results."

"The examples of modern SAT approaches to the problems of routing and the formal equivalence checking of digital circuits' descriptions within the technological mapping as a part of the FPGA design flow are considered."

"The algorithm of the detailed routing of the FPGA switching blocks using the satisfiability problem is developed and presented."

"The comparison of the developed SAT-based algorithm with the well-known routing algorithm Pathfinder is presented by the criteria of the elapsed time and the achieved degree of routed nets in the switching blocks."

"It is specified that the considered Boolean satisfiability methods for the routing problem are capable to prove the circuit's unroutability, unlike the Pathfinder algorithm whose results can only implicitly indicate it."

"The paper demonstrates that the application of a more efficient SAT solver significantly accelerates the work of the suggested detailed routing algorithm."

Introduction

"The Boolean satisfiability problem, SAT, allows us to find the substitution of the input variables that makes the value of the given Boolean formula to be true."

"Despite the fact that currently only algorithms for solving the Boolean satisfiability problem with an exponential complexity are known, in practice, many modern SAT solvers are able to find solutions to specific complex problems in a reasonable amount of time."

"The Boolean satisfiability problem is often solved for a formula in a conjunctive normal form (CNF, clause normal form)."

"For easier presentation, as a rule, SAT problems are written in the form of a system of Boolean constraints, each of which is a formula in the CNF."

"One of the key reasons for the continuous improvement of methods for Boolean satisfiability problems is a wide range of their practical applications in various fields of science and technology (artificial intelligence, bioinformatics, hardware and software verification, automatic theorem proving, etc)."

Boolean Satisfiability Methods for the Problems of Modern Computer-Aided Design Systems

"A significant part of the research on solving the routing problem within the FPGA design flow is devoted to the application of Boolean satisfiability methods."

"An optimization of the method for the formation of a finite Boolean system of a complete routing solution and a use of a more efficient SAT solver GRASP [18] for large problems made it possible to expand the scope of the proposed method for assessing the routability of the entire FPGA, not just its separate areas."

"Since one route unites a number of tracks and switching keys, the method allows you to form a simpler and shorter Boolean formula for a required routing solution."

"The considered works allow us to conclude that the idea of combining Boolean satisfiability methods with classical routing algorithms is viable."

"The considered applications of the Boolean satisfiability methods for FPGA routing problems allow us to draw the following conclusions."

Sat Based Approach to a Detailed Routing of Switching Blocks

"The SAT-based approach to a detailed routing of switching blocks is intended for the second stage of two-level routing of island-style FPGAs."

"The detailed routing procedure is performed inside the switching blocks, for which at the first (global) stage of routing a set of project nets routed through them is defined."

"The task of detailed routing is to find an uncongested solution for the multitude of all nets in the switching block."

"Let us consider the algorithm of a detailed routing [19] of the switching block."

"The main stage of a detailed routing using the SAT approach is the creation of a Boolean formula that fully describes the routing solution of the considered switching block."

Conclusions

"The result of the analysis of the recent research works showed that for problems of a design automation of microelectronics Boolean satisfiability methods act best in combination with auxiliary methods and techniques that allow slowing the growth of the dimension of the SAT problem being solved."

"SAT methods are suitable for situations where a proof of the presence or absence of a solution is required, as an example, for the estimation of a placement's routability."

A Convex Programming Solution for Gate-Sizing with Pipelining Constraints [20]

This is a machine-generated summary of:

Naidu, Srinath R.: A convex programming solution for gate-sizing with pipelining constraints [20]

Published in: Optimization and Engineering (2021)

Link to original: https://doi.org/10.1007/s11081-021-09616-0

If you want to cite the papers, please refer to the original.

For technical reasons we could not place the page where the original quote is coming from.

Abstract-Summary

"This paper proposes a novel geometric programming based formulation to solve a gate-sizing and retiming problem in the context of circuit optimization."

"The gate-sizing aspect of the problem involves continuous variables while the retiming problem involves the placement of registers in the circuit and can be naturally modeled using discrete variables."

"This is the first effort that solves this problem in a single optimization framework."

Introduction

"Gate-sizing is a well-studied problem that attempts to find the right sizes for the gates of the circuit, so that the minimum circuit delay can be achieved while meeting overall constraints on total area of the circuit, and the total power consumed."

"The work of Hurst et al. [21] states that the problem of combining gate-sizing and retiming in a joint formulation has not been well solved."

"The discrete aspect in this problem has to do with placement of flops so that the circuit is divided into two stages and the continuous problem has to do with the sizing of gates since gate sizes can take on continuous values."

"It must be noted here that there have been approaches in the past that have combined gate sizing and clock skew scheduling to optimize sequential circuit performance, for example Chuang et al. [22] and Satyamurthy et al. [23]."

Basic Formulation for Gate-Sizing

"We proceed to describe the approach of Boyd et al. [24] to the gate-sizing problem since it forms a basis of our work."

"The work of Boyd et al. [24] is a convex programming based approach that uses dynamic programming to minimize the arrival time at the output of a combinational circuit, subject to convex constraints on the total area and power of the circuit."

"The delay of a gate in the circuit is modeled using the Elmore delay model as a posynomial, which is an approach followed also by Fishburn and Dunlop [25]."

"If a gate is sized up, its area will go up but its delay will go down on account of its lower driving resistance."

Extending the Basic Formulation to Retiming

"There are an exponential number of possible register configurations corresponding to the number of cuts in an appropriate circuit graph representation, and it is impractical for one to enumerate all the possible configurations and solve gate-sizing problems for each cut."

"We seek to solve a single gate-sizing problem, possibly enhanced with additional variables, that will search through all possible register configurations and identify a bunch of promising register configurations that will help solve the pipelining problem."

"Our approach to enhance a gate-sizing formulation is to add a number of binary variables, one for each register position that will encode the presence or absence of a register at each location."

"In order to focus attention on the more promising register configurations, we use an objective function that biases the solution to the gate-sizing problem so as to favor such configurations."

Formulation Details

"The model above assumes that there is a single register placed at the output of a gate."

"This is made possible through the use of register variables, one for each signal line going from the output of a gate to the input of another gate."

"The objective function must favor a solution that places gates in partition 1 before moving on to partition 2."

"This solution is a frivolous solution in that it gives no insight as to where the true register front lies since all gates are placed equally in partitions 1 and 2."

"The first GP does not require very accurate gate-delay models since it comes up with a ranking of register positions which will not change too much with the use of more accurate gate-delay models, for example from SPICE."

Practical Solution to the Convex Program

"The Geometric Program as constructed above can in principle be solved by second-order methods such as the one employed by Boyd et al. [24] in their work."

"Second order methods build a Hessian matrix of size given by the number of variables in the convex program."

"First order methods use the gradient of the objective function to obtain the global minimum."

"It may be remarked in this context that the work of Joshi and Boyd [26] does solve a similar-looking GP to ours using second-order methods."

"In our case the constraints are non-linear and we need to find another method to solve the sub-problem."

"We find in practice that a modest number of points generated in the feasible region are sufficient to solve the sub-problems and the global problem accurately enough for the method to report the best cut."

Generating Points in the Feasible Region

"The heart of the first-order method to solve the convex program is the generation of points within the feasible region of the constraint space."

"Given the requirement to obtain a representative set of points we take special care to generate points within the feasible region."

"Sizes for gates in the circuit can be assigned according to various criteria."

"We can draw its delay from an exponential distribution around the minimum delay and assign its size based on the value of the delay."

"When the gate-sizes and partition variables are assigned, we perform a back-calculation from primary outputs back to primary inputs to see if we meet the arrival time constraints."

Projecting Feasible Points to the Convex Region Boundary

"We need an efficient way to project a feasible point onto the boundary."

"One obvious way is to proceed in the given direction and check feasibility of the shifted point against all constraints."

"We can find a point on the line joining these two points that falls on the boundary of the feasible region using binary search."

"In order to determine feasibility of a given point, we would need to check if all constraints are feasible for the point."

"As the algorithm keeps checking points closer and closer to the boundary for feasibility, it declares success when the critical list of constraints at a point becomes empty."

"Using binary search we identify the point C that lies on the boundary."

Number of Iterations

"The trick is to project the feasible points onto the boundary for the first few iterations and find in each iteration the best sample, i.e., the sample that minimizes the linearized cost function for that iteration."

"Having obtained a set of best sample points for the first few iterations, we then use only these samples in subsequent iterations instead of the whole feasible point set."

"This makes sense since the cost function vector starts to stabilize after the first few iterations, so that it is likely that only the same small number of the sample points in the feasible set will decide optimality in subsequent iterations."

Finding a Feasible Cut

"Having obtained a list of weights for the different locations, we sort the weights in descending order and cycle through various "promising cuts" in order to find a valid feasible cut."

"In order to make the concept outlined above workable, we need to first decide how to calculate weights for various wires in order to decide their importance from the point of view of placing a register."

"We sort the computed weights in descending order to create a list of candidates for the placement of registers."

"Our algorithm is based on the intuition that the weights on the wires reflect their true importance so that a sorted list of weights (in descending order) gives us a list of weights that are close to a viable register placement."

"Among all remaining positions, the procedure picks one with the highest weight according to the given list and inserts a register at the new location."

Gate-Sizing for a Cut

"In the context of our problem, gate-sizing is essentially a convex projection problem in that we seek to project an infeasible solution onto the convex feasible region."

"We approximate the given true convex feasible region by a convex polytope, and then use a reasonably fast projective algorithm to project the given initial solution on to the convex polytope."

"We can use this projected point as the starting point of a gate-sizing solution."

"In order to find a tight polytope we need to take tangents to the bounding surfaces of the feasible region at points that actually constitute the boundary of the feasible region."

"The process of finding a tight polytope requires us to resort to the random-search based gate-sizing procedure described above in an effort to find gate-sizes that minimize the maximum arrival time."

Computational Complexity

"Having built clausal sub-circuits which represent the clauses in the 3-SAT formula we proceed to build the rest of the circuit to simulate the assignment of Boolean variables across clauses so that a given variable in different clauses is assigned the same value."

"For a given Boolean assignment every clausal gate will see a particular load and for a large enough total area budget we can assign sizes to the gates that make up the clausal gate so that the delay of each clausal gate is the smallest possible when the clausal gate drives the given load."

"To ensure this happens we consider each clause type, and determine the resistance needed for the driving gate of the clausal sub-circuit corresponding to each clause type and assign resistances to the different clause types as follows: The above assignment to the driving resistances of the different clausal gates ensures although different clausal sub-circuits see different loads when they are given an unsatisfying assignment, the optimal delay seen by all these clausal sub-circuits is the same."

Experimental Results

"We obtain the solution to the convex programming problem and compute the weights for the wires according to each measure."

"Each time we use a particular measure we perturb the ranking slightly to ensure that we explore a configuration close to the order suggested by the weight measure but not exactly the same as the one suggested by the weight measure."

"During the round-robin phase of biased search we calculate the rank of each weight measure as a function of the quality of the configurations produced by the particular weight measure and its perturbations."

"We use the round-robin phase of biased search for 48 configurations and then use sampled weight measures for the remaining configurations."

"The random exploration of configurations takes place via a Markov chain random walk on a configuration graph where each configuration is connected by an edge to a configuration that is obtained through the motion of registers either from all the inputs of a gate to its outputs or vice-versa."

Conclusion and Further Work

"We presented a method for solving the gate-sizing-cum-retiming problem which is an optimization problem having both discrete and continuous components."

"There are a number of parameters in our solution that need to be tuned; for example, the solution to the convex programming problem requires us to find a number of feasible solutions in the constraint space and project them to the boundary of the feasible region—the method we have chosen to generate feasible solutions is a heuristic which performs well in practice but there is a possibility for further investigation in this area."

"There is scope for investigating how our method can be applied to more realistic delay models, not just those which are compatible with Geometric Programming techniques."

"In the light of the work of Joshi and Boyd [26], one avenue of further research could be to develop faster second-order methods to solve the GPs proposed in our paper."

A Survey of Algorithmic Methods in IC Reverse Engineering [27]

This is a machine-generated summary of:

Azriel, Leonid; Speith, Julian; Albartus, Nils; Ginosar, Ran; Mendelson, Avi; Paar, Christof: A survey of algorithmic methods in IC reverse engineering [27]

Published in: Journal of Cryptographic Engineering (2021)

Link to original: https://doi.org/10.1007/s13389-021-00268-5

If you want to cite the papers, please refer to the original.

For technical reasons we could not place the page where the original quote is coming from.

Abstract-Summary

"The discipline of reverse engineering integrated circuits (ICs) is as old as the technology itself."

"The process of IC reverse engineering comprises two steps: netlist extraction and specification discovery."

"While the process of netlist extraction is rather well understood and established techniques exist throughout the industry, specification discovery still presents researchers with a plurality of open questions."

"We present a survey of the state of the art in IC reverse engineering while focusing on the specification discovery phase."

"We observe that the state of research on algorithmic methods for specification discovery suffers from the lack of a uniform evaluation approach."

"We point out the urgent need to develop common research infrastructure, benchmarks, and evaluation metrics."

Introduction

"Additional applications of IC reverse engineering include the detection of counterfeit devices, failure analysis, and monitoring of semiconductor suppliers."

"On the technical side, reverse engineering of Integrated Circuits (ICs) is a complex process that involves multiple disciplines and skills."

"That in this work, the term specification is used in a broader context, generally referring to the objective of the reverse engineering process."

"During the first phase, netlist extraction, the IC is analyzed to generate a gate-level netlist description."

"Our work mainly focuses on the second phase, specification discovery, which takes the gate-level netlist as an input and results in a partial or full understanding of the IC's functionality."

"To address that problem, we first need to answer the question of how to represent the result of the reverse engineering process."

Netlist Extraction

"Delayering and Imaging The layers of the chip are removed in interleaving with an image acquisition step to retrieve images of every layer."

"A special software is used to extract the gate-level netlist by identifying standard cells and their interconnections in the metal layers [28]."

"A bitstream is a different, usually proprietary, representation of the gate-level netlist implemented on the device."

"The process of retrieving the netlist from an FPGA can be split into the steps of extracting the bitstream from memory, understanding the bitstream file format, and converting the bitstream into a gate-level netlist."

"Since differences in the bitstream directly relate to the changes introduced in the altered design, the attacker can deduce a database that maps individual bits of the bitstream to the corresponding elements in the netlist, as well as their configuration."

"This database can subsequently be used by an attacker to convert the bitstream into the desired gate-level netlist."

Specification Discovery: Fundamental Algorithms

"One such algorithm, SubGemini [29], identifies logic gates within a transistor-level netlist by solving a subgraph isomorphism problem."

"Understanding circuits of millions of gates first requires partitioning into smaller subcircuits to allow for efficient understanding of high-level functionality."

"In [30], a pure graph-based algorithm that does not use any functional information splits the circuit into densely connected subgraphs."

"Discriminative labeling, the vertices in both the objective function (the investigated circuit) and the model (library components) are labeled using a recursive labeling algorithm that considers each vertex's surroundings."

"The reverse engineering task then turns into a problem of matching functional blocks in the analyzed netlist with a library of pre-built HLDs."

"The challenge is, given a circuit with inputs comprising of words and other (control) signals, to find assignments of the control signals such that the word-level functionality of the circuit will match one of the templates."

Specification Discovery: Putting It All Together

"It combines functional and structural methods to identify and extract the FSM circuitry and subsequently reconstruct the state graph."

"The RELIC tool [31] performs a topological analysis on the netlist to find similar fan-in cones of cells and is aiming to identify state registers."

"The tool is evaluated on a number of small benchmark circuits and shows 80–100% accuracy in the detection of control registers."

"Transitions between the states are identified by evaluating the combinational logic feeding the state registers."

"Multi-bit registers are identified using the bitslice aggregation algorithm by combining a set of registers that have a common write control."

"Processing phase Each of the metrics applied in this phase has to follow a fixed set of rules before a register can be created: (1) FFs need to share common clock and control signals and (2) be in the same register stage."

Reverse Engineering Tools and Frameworks

"A number of tools from academia and commercial industries have been developed to aid in the reverse engineering process."

"The Hardware Analyzer (HAL) [32–34] is a comprehensive reverse engineering and manipulation framework for gate-level netlist."

"HAL is designed to aid in the extraction of high-level information from gate-level netlists."

"The tool receives images of the layers of an IC as well as a standard cell library and subsequently outputs a gate-level netlist."

"They develop an automatic reverse engineering tool called ChipJuice."

"This tool processes layer images of ICs, identifies wires and devices, assembles them to standard cells, and finally generates the corresponding gate-level netlist [35]."

"There is a limited tool support in throughout the industry and in academia for IC reverse engineering."

"The few tools that exist are mainly dealing with the first phase of the reversing process, netlist extraction."

Summary, Challenges, Open Questions, and Future Directions

"The key to perform such attacks successfully is to use internal information that can be obtained by extracting the inner structure of the chip using hardware reverse engineering."

"Meade and others [36] brought to attention that most gate-level netlist reverse engineering techniques lack proper evaluation."

"The NMI metric is a good candidate to adopt for at least some reverse engineering methods, such as netlist partitioning and separation of datapath and control logic."

"Unlike classical problems that show promising results using machine learning algorithms, such as pattern recognition in images, the size of the training dataset is strongly limited in the case of netlist reverse engineering."

"The classification of existing work on algorithmic methods for IC reversing presented in this paper may serve as a framework for evaluation of countermeasures against different kinds of reverse engineering attacks."

Gate Sizing Using Geometric Programming [37]

This is a machine-generated summary of:

Posser, Gracieli; Flach, Guilherme; Wilke, Gustavo; Reis, Ricardo: Gate sizing using geometric programming [37]

Published in: Analog Integrated Circuits and Signal Processing (2012)

Link to original: https://doi.org/10.1007/s10470-012-9943-3

If you want to cite the papers, please refer to the original.

For technical reasons we could not place the page where the original quote is coming from.

Abstract-Summary

"A two-step transistor sizing optimization method based on geometric programming for delay/area minimization is presented."

"Elmore delay is minimized using only minimum and maximum transistor size constraints."

"The minimized delay found in the previous step is used as a constraint for area minimization."

"By relaxing the minimized delay, one may further reduce area with small delay penalty."

"Gate sizing and transistor sizing are performed considering delay minimization."

"Gate sizing is able to reduce delay in 21%, on average, for the same area and power values of the sizing provided by standard-cells library."

"The transistor sizing is executed and delay can be reduced in 40.4% and power consumption in 2.9%, on average, compared to gate sizing."

"Gate sizing optimizing area is executed considering a delay constraint."

"Considering the minimum delay, area (power) is reduced in 28.2%, on average."

"Relaxing delay by just 1%, area (power) is reduced in 41.7% and the EDG metric is 41.7."

"Area can be reduced in 51%, on average, relaxing delay by 5% and EDG metric is 10.2."

Introduction

"One can make a circuit faster, occupy less area or consume less power by sizing its transistors properly."

"The goal of a transistor sizing (TS) optimization is to minimized delay subject to area constraints or similarly to minimize area (or power) subject to a timing constraints."

"Gate sizing (GS) is a special case of TS where transistors composing a gate are tied to a same scale factor."

"The tool can be used along with an automatic cell layout generation tool taking advantage of "continuous" GS avoiding the rounding-off problems of traditional GS approaches; a transistor and GS tool configurable to several manufacturing technology by changing the technology parameters; a comparison between GS and TS taking account the tradeoff between number of variables (execution time) and circuit total delay reduction; a study about delay-area (power) tradeoff considering GS."

Related Work

"TILOS attempts to meet the delay constraint by increasing the size of as few transistors as possible."

"Delay is modeled as posynomial functions of the transistor sizes, creating a GP problem that is transformed into a convex program, which is optimally solved."

"Transistor sizing presented in [38] sizes the transistors of CMOS circuits to minimize power consumption while the delay constraint is met without inserting changes in circuit structure and number of gates."

"Non-linear programming is used in [39, 40]. Reference [26] is concerned about scalability of the circuit, i.e., the execution time needed to size large circuits. References [41, 42] use power, delay and area analytical models."

"The gate delays are estimated by a linear function on the scale factors, where the delay of each different gate with size one is calculated and the delay of other gates of the same type is linear at this value."

"For each speed, an optimization problem is solved, finding new sizes for the gates such that the circuit satisfies the speed goal while dissipating minimal power."

Geometric Programming

"The term geometric program was introduced in 1967 [43] and describes an optimization problem of the form where f_i are posynomial functions, g_i are monomials, and x_i are the optimization variables."

"A posynomial function is defined as the sum of one or more monomial functions."

"The equality constraints must be in monomial form and equal to one."

"The inequality constraints must be posynomials and less than or equal to one [24]."

Problem Formulation

"Two different formulations are outlined: (1) delay minimization subject to an area constraint (2) and area minimization subject to a delay constraint."

"X_{min} and X_{max} are the minimum and maximum gate sizes (GS) or the minimum and maximum transistor widths (TS)."

"In area minimization the objective function becomes area and delay becomes a constraint."

Tool Development

"The inputs of the tool are: a SPICE netlist, load capacitance, maximum input capacitance and area or delay restriction."

"Capacitance and resistance values used to calculate delay and power are obtained by SPICE simulations considering PMOS and NMOS transistors."

"The transistor capacitances are proportional to the transistor width."

"The driving resistance is considered inversely proportional to the transistor width."

"Area is the sum of the width W_i of each transistor i, where n is the number of transistors of the circuit: 6."

"Switching power is calculated considering all capacitances of the circuit: load capacitance (C_{load}) and input capacitances (Cin)."

Tool Verification

"The delay and power values from liberty file obtained by the characterization are used to compare the cells sized by different methods."

"The worst-case delay values are compared for different sizing methods for each selected gate."

"It can be seen that the GS using GP has lower delay and lower power consumption than other methods (TSS, DS, SC and PNR)."

"The lower gain in delay is obtained when GP-based sizing is compared to P/N--based sizing (PNR), where the gain is only 0.6% because the transistor sizes are nearly equal."

"The lower delay gain is obtained when GP-based sizing is compared to drive strength-based sizing (DS)."

"The sizing tool using GP always achieves the lowest delay or the same delay than other methods."

"The same delay is obtained when the sizing obtained by GP is the same given by an other method."

Transistor Sizing Versus Gate Sizing

"In TS, for each transistor of the circuit a variable is set representing the transistor width."

"Reduction (R) values represent how much the GS using GP reduces power, timing and area compared with the sizes available in the standard cell library."

"The number of auxiliary variables used in both, GS and TS, is the same for the same circuit."

"The total number of variables for the GS is the sum of the number of gates and the number of auxiliary variables, for example, the number of variables to the C432 circuit is 528 (184 + 344)."

"The total number of variables for the TS is given by the sum of the number of transistors in the circuit and the number of auxiliary variables, in the C432 circuit, the number of variables in TS is 1,010 (666 + 344)."

Delay Minimization Followed by Area Minimization

"Area and power values given by delay minimization (Delay Min.) followed by area minimization (Area Min.) are shown."

"Area reduction achieves 28.2%, on average, and 27.3% is the reduction on power consumption, considering the same delay value given by delay minimization, i.e., keeping the optimum delay."

"This improvement is possible when there are multiple optimal points that minimize the delay, given an area budget."

"The area minimization problem is able to find the minimum area considering the minimal circuit delay."

Energy-Delay Tradeoff Results

"The delay relaxed by 5% is used as constraint in area (power) minimization."

"A 41.7% reduction in area (power) can be achieved by relaxing delay by 1%, compared to delay minimization."

"The tradeoff between area (power) and delay, EDG, relaxing delay by 1% has the same average value, 41.7."

"Considering delay relaxed by 5%, area reduces 51%, on average, and EDG is 10.2, on average."

"Relaxing the optimum delay slightly can provide significant circuit area (power) reduction."

Conclusions and Future Works

"The sizing is compared to other four methods to minimize delay subject to an area constraint that is equal for all methods."

"Using the sizing tool developed in this work, delay can be reduced in 21%, on average, for the same area and power values of the sizing provided by standard-cells library, considering delay minimization."

"TS reduced delay in 40.4% and power consumption in 2.9%, on average, compared to the results from GS."

"Delay minimization followed by area minimization considering the minimum delay as constraint was able to reduce area (power) in 28.2%, on average."

"Relaxing delay by 1% reduces 41.7% of area, on average, compared to the area given by delay minimization and the EDG (Energy/delay gain) is 41.7, on average."

"When delay is relaxed by 5%, area is reduced in 51%, on average, and EDG is 10.2, on average."

Latency-Sensitive High-Level Synthesis for Multiple Word-Length DSP Design [44]

This is a machine-generated summary of:

Le Gal, Bertrand; Casseau, Emmanuel: Latency-Sensitive High-Level Synthesis for Multiple Word-Length DSP Design [44]

Published in: EURASIP Journal on Advances in Signal Processing (2011)

Link to original: https://doi.org/10.1155/2011/927670

If you want to cite the papers, please refer to the original.

For technical reasons we could not place the page where the original quote is coming from.

Abstract-Summary

"High-level synthesis (HLS) currently seems to be an interesting process to reduce the design time substantially."

"Conventional HLS techniques usually focus on uniform-width resources according to the worst-case data requirements, that is, the largest word length."

"An HLS process that takes care of the delay of the operators according to the data width is presented."

"Experimental results show that our approach achieves significant design latency saving or area decrease compared to a conventional synthesis."

References

1. Chowdhury P, Guin U, Singh AD, Agrawal VD (2021) Estimating operational age of an integrated circuit. J Electron Testing 37:25–40. https://doi.org/10.1007/s10836-021-05927-3
2. Reddy V, Krishnan AT, Marshall A, Rodriguez J, Natarajan S, Rost T, Krishnan S (2005) Impact of negative bias temperature instability on digital circuit reliability. Microelectron Reliab 25(1):31–38
3. Schroder DK, Babcock JA (2003) Negative bias temperature instability: Road to cross in deep submicron silicon semiconductor manufacturing. J Appl Phys 94(1):1–18
4. Zhang X, Xiao K, Tehranipoor M (2012) Path-delay fingerprinting for identification of recovered ICs. In: Proceedings of the IEEE International Symposium on Defect and Fault Tolerance in VLSI and Nanotechnology Systems (DFT). ACM, pp 13–18
5. Zheng Y, Wang X, Bhunia S (2015) SACCI: Scan-based characterization through clock phase sweep for counterfeit chip detection. IEEE Trans Very Large Scale Integr Syst 23(5):831–841
6. Saneyoshi E, Nose K, Mizuno M (2010) A precise-tracking NBTI-degradation monitor independent of NBTI recovery effect. In: IEEE International Solid-State Circuits Conference Digest of Technical Papers (ISSCC). IEEE, pp 192–193
7. Zhang X, Tehranipoor M (2014) Design of on-chip lightweight sensors for effective detection of recycled ICs. IEEE Trans Very Large Scale Integr Syst 22(5):1016–1029
8. Zhang X, Tuzzio N, Tehranipoor M (2012) Identification of recovered ICs using fingerprints from a light-weight on-chip sensor. In: Proceedings of the IEEE-ACM Design Automation Conference. IEEE
9. He K, Huang X, Tan SXD (2015) EM-based on-chip aging sensor for detection and prevention of counterfeit and recycled ICs. In: Proceedings of the IEEE/ACM International Conference on Computer-Aided Design (ICCAD). IEEE, pp 146–151
10. Alam M, Chowdhury S, Tehranipoor M, Guin U (2018) Robust, low-cost, and accurate detection of recycled ICs using digital signatures. In: Proceedings of the IEEE International Symposium on Hardware Oriented Security and Trust (HOST). IEEE
11. Wang W, Reddy V, Yang B, Balakrishnan V, Krishnan S, Cao Y (2008) Statistical prediction of circuit aging under process variations. In: Proceedings of the Custom Integrated Circuits Conference (CICC). IEEE, pp 13–16
12. Brglez F, Fujiwara H (June 1985) A neutral netlist of 10 combinational benchmark circuits and a targeted translator in FORTRAN, special session on ATPG and fault simulation. In: Proceedings of 1985 IEEE International Symposium on Circuits and Systems (ISCAS'85). IEEE, pp 663–698
13. Sayil S, Lamichhane S, Sayil K (2020) Crosstalk noise mitigation using a transmission gate with varied gate bias. Analog Integr Circ Sig Process 105:183–190. https://doi.org/10.1007/s10470-020-01706-w
14. Becer MR et al (2002) Analysis of noise avoidance techniques in DSM interconnects using a complete crosstalk noise model. In: Proceedings 2002 Design, Automation and Test in Europe Conference and Exhibition, Paris, France. IEEE, pp 456–464
15. Sayil S (2016) Soft error mechanisms, modeling and mitigation. Springer, Berlin
16. Ferlet-Cavrois V, Massengill LW, Gouker P (2013) Single event transients in digital CMOS-a review. IEEE Trans Nuclear Sci 60(3):1767–1790
17. Zapletina MA, Zhukov DV, Gavrilov SV (2021) Boolean satisfiability methods for modern computer-aided design problems in microelectronics. Russ Microelectron 50:516–522. https://doi.org/10.1134/s1063739721070131
18. Marques Silva JP, Sakallah K (1997) Grasp—a new search algorithm for satisfiability. In: Proceedings of the 1996 IEEE/ACM International Conference on Computer-Aided Design ICCAD'96, USA. IEEE Computer Society, pp 220–227
19. Zhukov DV, Zheleznikov DA, Zapletina MA (2020) The iterative SAT based approach to detailed routing for reconfigurable system-on-a-chip. In: Proceedings of the 2020 IEEE Conference of Russian Young Researchers in Electrical and Electronic Engineering EIConRus, pp 1905–1910

20. Naidu SR (2021) A convex programming solution for gate-sizing with pipelining constraints. Optim Eng 23:947–982. https://doi.org/10.1007/s11081-021-09616-0
21. Hurst A, Mischenko A, Brayton RK (2007) Minimizing implementation costs with end-to-end retiming. In: International workshop on logic sysnthesis. The Association for Computing Machinery, New York
22. Chuang W, Sapatnekar SS, Hajj IN (1993) A unified algorithm for gate sizing and clock skew optimization to minimize sequential circuit area. In: Proceedings of the International conference on computer-aided design. IEEE
23. Satyamurthy H, Sapatnekar SS, Fishburn JP (1998) Speeding up pipelined circuits through a combination of gate sizing and clock skew optimization. IEEE Trans Comput-Aided Des Integr Circuits Syst 17(2):173–182
24. Boyd S, Kim S-J, Patil D, Horowitz M (2005) Digital circuit optimization via geometric programming. Oper Res 53(6):899–932
25. Fishburn J, Dunlop A (1985) Tilos: a posynomial programming approach to transistor sizing. In: IEEE international conference on computer-aided design. IEEE, pp 326–328
26. Joshi S, Boyd S (2008) An efficient method for large-scale gate sizing. IEEE Trans Circuits Syst I 55:2760–2773
27. Azriel L, Speith J, Albartus N, Ginosar R, Mendelson A, Paar C (2021) A survey of algorithmic methods in IC reverse engineering. J Cryptographic Eng. https://doi.org/10.1007/s13389-021-00268-5
28. Quadir SE, Chen J, Forte D, Asadizanjani N, Shahbazmohamadi S, Wang L, Chandy J, Tehranipoor M (2016) A survey on chip to system reverse engineering. ACM J Emerg Technol Comput Syst 13(1):1–34
29. Ohlrich M, Ebeling C, Ginting E, Sather L (1993) SubGemini: identifying subcircuits using a fast subgraph isomorphism algorithm. In: Proceedings of the 30th International on Design Automation Conference—DAC'93. ACM Press, New York, pp 31–37
30. Couch J, Reilly E, Schuyler M, Barrett B (2016) Functional block identification in circuit design recovery. In: 2016 IEEE International Symposium on Hardware Oriented Security and Trust (HOST). IEEE, pp 75–78
31. Meade T, Jin Y, Tehranipoor M, Zhang S (2016) Gate-level netlist reverse engineering for hardware security: control logic register identification. In: 2016 IEEE International Symposium on Circuits and Systems (ISCAS). IEEE, pp 1334–1337
32. Chair for Embedded Security: HAL—The Hardware Analyzer (2019) https://github.com/emsec/hal
33. Fyrbiak M, Wallat S, Déchelotte J, Albartus N, Böcker S, Tessier R, Paar C (2018) On the difficulty of FSM-based hardware obfuscation. IACR Trans Cryptogr Hardw Embed Syst 2018:293–330
34. Wallat S, Albartus N, Becker S, Hoffmann M, Ender M, Fyrbiak M, Drees A, Maaen S, Paar C (2019) Highway to HAL: Open-Sourcing the first extendable gate-level netlist reverse engineering framework. In: ACM International Conference on Computing Frontiers 2019, CF 2019—Proceedings. ACM, pp 392–397
35. Thomas O, Sarl T, Nedospasov D (2015) On the impact of automating the IC analysis process. Tech Rep 1–10
36. Meade T, Shamsi K, Le T, Di J, Zhang S, Jin Y (2018) The old frontier of reverse engineering: netlist partitioning. J Hardw Syst Secur 2(3):201–213
37. Posser G, Flach G, Wilke G, Reis R (2012) Gate sizing using geometric programming. Analog Integr Circ Sig Process 73:831–840. https://doi.org/10.1007/s10470-012-9943-3
38. Borah M, Owens R, Irwin M (1996) Transistor sizing for low power cmos circuits. IEEE Trans Comput Aided Design Integr Circuits Syst 15(6):665–671
39. Mahalingam V, Ranganathan N (2005) A nonlinear programming based power optimization methodology for gate sizing and voltage selection. In: ISVLSI 2005: IEEE computer society annual symposium on VLSI. IEEE, Tampa, pp 180–185
40. Sapatnekar SS, Chuang W (2000) Power-delay optimization in gate sizing. ACM Trans Design Automat Electron Syst 5(1):98–114

41. Hoppe B, Neuendorf G, Schimitt-Landsiedel D, Specks W (1990) Optimization of high-speed cmos logic circuits with analytical models for signal delay, chip area and dynamic power dissipation. IEEE Trans Comput Aided Design Integr Circuits Syst 9(3):236–246
42. Borah M, Owens RM, Irwin MJ (1995) Transistor sizing for minimizing power consumption of CMOS circuits under delay constraint. In: 1995 International symposium on low power design. Dana Point, pp 167–172
43. Duffin RJ, Peterson EL, Zener C (1967) Geometric programming-theory and application. Wiley, New York
44. Le Gal B, Casseau E (2011) Latency-sensitive high-level synthesis for multiple word-length DSP design. EURASIP J Adv Signal Process 2011:927670. https://doi.org/10.1155/2011/927670

Chapter 3
Optimization Methods in Physical Design

Introduction by the Editor

Physical design (PD) is a critical step in VLSI chip design. PD is placing design block in netlist in a layout which will mimic silicon placement. PD consists of many steps to convert a netlist to a physical layout. As more and more logic is packed in smaller technology nodes, it has become critical to optimize every step in PD flow. This chapter elaborates on various methods that can be used for optimizing PD flow. It discusses many algorithms which help with PD flow. Heterogenous FPGAs—PhetDP are key algorithms used to optimize whole flow. Algorithms like Genetic algorithms-based approach to solve Clustered Steiner tree problem are applicable here. Accelerating algorithms such as accelerated computation of Steiner trees on GPUs, Goal oriented shortest path search and Fast DSE of reconfigurable accelerator systems are also elaborated in this chapter. Thermal hotspots are aware of multi-level routing and heuristic search for MEC, optimized placement solution for mixed size designs and circuit partitioning algorithms, partition models and algorithm parallelization are also discussed in this chapter. Authors have surveyed and discussed all these algorithms in detail.

Ultra Scale heterogenous FPGAs utilizes architectural information of the FPGA to guide the placement from the beginning. Cluster selection and net selection are the processes which follow. This framework has a fixed-block based cluster formation step, separation table formation step and cell-mapping step. Later two steps are specifically used to get a feasible global placement solution by considering an estimated localization and area congestion. A new legalization technique is being used to reduce effort by using delayed packing and cell mapping to BELs. Wirelength reduction is achieved by using congestion free placement of blocks.

Steiner Tree Problem (STP) computes a minimum weighted graph such as extracting a subset of nodes called terminals. STP is a NP-hard problem. Polynomial-time approximation techniques help solve STP. Kou, Markowsky and Berman

A. S. Shastri (ed.), *Optimization Methods in VLSI Design*,
https://doi.org/10.1007/978-981-95-2457-0_3

(KMB) algorithm provide a two-step approximation to STP. Loosely modelled KMB is inefficient for large graphs. The polynomial-time approximation algorithm enhances the practical applicability of the KMB algorithm by leveraging GPU parallelization, thereby decreasing its execution time on real-world graphs. This parallelization has many challenges as the compute doesn't understand similarities in data. To solve this problem, GPU based compute approach is used for STP with KMB.

Routing in VLSI is to connect pins to pins using a net or group of nets. Other constraints like parasitic, timing convergence, etc. also needs to be taken into consideration. Pins are considered as vertices of a graph. Each net consists of at least one input and one output which needs to be connected using STP. Far-away nets should be disjoint vertex in STP. These graph connections are replaced with routing tracks. Any two routing tracks on adjacent layers are orthogonal to each other. These are metal connections called vias. STP helps with optimization of global routing considering power, timing convergence and area minimization.

With IOT and AI coming on edge devices, real-time computing has become very important in mobile edge computing (MEC) environment. Real-time computing calls for strict minimization of delay. In routing, overflow, wirelength and runtime are the main evaluation standards. Optimization of the same is becoming challenging with complex design blocks. Fast lookup table (FLUTE) algorithm builds a topology of each net with many Steiner points. Prim's and Divide and Conquer algorithm build a topology that avoids creating many Steiner points which with ease of computation. Once connection and routing are done, congestion area identification is used to Rip-up and reroute (R&R) connections to reduced congestion. A custom search operation is used which considers both congestion and wire length of the Nets and optimises it.

In design blocks which contains complex network application. There might be notes, which are partitioned in different local clusters, but are needed to function together to perform the job intended. This basically is an additional constraint which is added to Steiner tree problem. This is also called as clustered Steiner tree problem. A genetic algorithm which is a shortest path, can help resolve this cluster strainer problem. This is basically an approach which uses permutation to individually connect the Nets and reduce the dimensionality of the cluster. This is applicable for Let's in both Euclidean and round Euclidean cases.

When we have design Blogs on FPGAs, we can have design blocks with different sizes and shapes. An optimised placement approach is used for such cases. This is a heuristic approach which considers routability of cells, clusters and pins. This ensures that the layout is manageable and can be connected without any congestion. Method also includes the process of unique legalisation used in circle rolling in block sliding techniques. Detailed placement base can pass out the cells responsible for congestion and make them layout friendly.

While Partitioning, a circuit is recursively divided into sub circuits so that each can be independently designed to save cost and connect the whole circuit as chiplets.

Is the first step of the backend flow in VLSI design. Steps like floor planning, placement pin assignment, pin connectivity, routing, etc. depends on this step's outcome. The partition better is the overall quality of the final layout. Novel partitioning algorithm based on the concept of satin Bowerbird optimization (SBO) picks a random partition, places it in the layout and look for all the design blocks which are compatible with it and place those blocks with it Using fitness value evaluation. If there is a need of any other placement changes, this can be used as a group to migrate to another location of the layout. This improves cost.

The recent VLSI trends. Reconfigurable hardware accelerators are becoming a good choice to meet Performance goals. Hardware accelerator is a hardware block which specialises in specific application and is optimised on Area, power and timing. Space exploration through hardware synthesis extremely time consuming. This can be modified using machine learning algorithms. Once design is given as an input to machine learning algorithm, it can Help predict aspects like performance and Timing and area. Ensemble Learning Is a well implemented approach since it can use multiple models and combine their prediction to give a final score. This simplifies an accelerate design space exploration process for general purpose processors with Hardware accelerators.

Three Dimensional circuits have become most popular technology as it helps pack more and more transistor which means more and more logic in the Confined space. New dimensions adding to routing. Wireline can be huge and routing parts when create congestion. it is essential to reduce wire length for optimal performance of the chip. Reducing wireless also helps with power dissipation. These multilayer connections can generate heat. Genetic algorithm Is used to estimate These hotspots. Hot spots identified; we can work on minimising the wire length between those components.

Machine Generated Summaries

Disclaimer: The summaries in this chapter were generated from Springer Nature publications using extractive AI auto-summarization: An extraction-based summarizer aims to identify the most important sentences of a text using an algorithm and uses those original sentences to create the auto-summary (unlike generative AI). As the constituted sentences are machine selected, they may not fully reflect the body of the work, so we strongly advise that the original content is read and cited. The auto generated summaries were curated by the editor to meet Springer Nature publication standards. To cite this content, please refer to the original papers.

Machine generated keywords: placement, graph, wirelength, steiner tree, steiner, global, cluster, congestion, cell, global placement, vertex, routing, FPGAs, layer, pin

PHetDP: A Placement Algorithm for Heterogeneous FPGAs with Delayed Packing [1]

This is a machine-generated summary of:

Purkayastha, Sharbani; Mukherjee, Shyamapada: PHetDP: A Placement Algorithm for Heterogeneous FPGAs with Delayed Packing [1]

Published in: Circuits, Systems, and Signal Processing (2022)

Link to original: https://doi.org/10.1007/s00034-022-02159-4

If you want to cite the papers, please refer to the original.

For technical reasons we could not place the page where the original quote is coming from.

Abstract-Summary

"A new wirelength-driven placement model for UltraScale heterogeneous FPGAs have been developed and presented."

"A detailed placement phase is explored to refine the placement to achieve congestion free legal placement of cells with minimized wirelength."

"Experimental results have shown that the proposed algorithm achieves better results than the state-of-the-art FPGAs placement algorithms in terms of wirelength without using any expensive optimization framework."

Introduction

"This paper presents a wirelength-driven placement approach which considers the architecture of the UltraScale heterogeneous FPGAs to guide the placement algorithm from the initial placement stage itself."

"The global placement framework of the proposed algorithm includes a lookahead legalization-based global placement which facilitates the placement of a circuit with minimized total wirelength and balanced cell density."

"The last and final step, detailed routing, refines the placement results by minimum cell movement which satisfies all the design rules, wirelength and congestion minimization."

"The main contributions of the paper are summarized as follows: A new fixed cell and white space-based initial placement for wirelength-driven placer has been presented."

"The proposed algorithm introduces a novel segregation table-based cell spreading approach for improving the effectiveness of global placement, rough legalization and area congestion."

"A delayed packing-based legalization has been introduced for cell mapping to BELs in order to reduce legalization cost and keeps wirelength minimum."

Literature Review

"The state-of-the-art placement approaches can be categorized into the following categories: simulated annealing, partition-based, force-directed, and analytical approach."

"A recursive partitioning-based congestion-driven placement approach is presented in [2] which requires only a single run of the placement algorithm."

"Ripple 2.0 [3] is a routability-driven placement algorithm which considers pin density and routing path based cell inflation."

"The proposed algorithm in the paper integrates global placement (Fastplace) which is force-directed, followed by detailed placement (FastDp), and routing (FastRoute) for finding a routable solution."

"A deep learning framework has been proposed in [4] that predicts routability with the help of CNN model to predict routability of a placement solution without conducting early global routing."

"The work in [5] has targeted timing requirement of clock-aware architecture with focus on detailed placement with machine learning-based routability prediction model."

Target Architecture

"The target architecture is consisting of logic elements like lookup tables (LUTs), flip-flops (FFs), RAMs, DSPs, and IOs."

"Each logic element can be put into a SITE of its own type."

"SLICEs are the SITEs for configurable logic blocks (CLBs)."

"RAM, DSP, and I/O can be put into its corresponding SITE type."

"A SLICE can accommodate 16 LUTs and 16 FFs, while SITEs for DSP and RAM can have only one logic element of its type."

"Several LUTs and FFs together form a basic logic element (BLE), and then several BLEs are grouped into a CLB."

Proposed Method

"After obtaining an initial placement, it performs cluster selection and net selection in order to identify the densely connected cells."

"In the proposed approach, a legalization-based cell mapping to BELs is applied on the placement result after global placement."

"In this phase, initially global bin locations are decided for the nets and their associated cells and further site locations are searched for global placement of the cells."

"The proposed step takes advantage of the fixed blocks of a circuit to obtain the initial placement of cells in the nets having fixed cell attached."

"Based on the capacity of a placement site for various cell types and legalization rules for a specific cell type, grouped cells may be split into more subgroups."

"The detailed placement finds suitable legal placement of these cells along with movement of other LUTs and FFs to remove congestion in some areas."

Results and Discussion

"Run time value also shows that the propose global placement performs reasonably well in terms of run time."

"The column head "LegSat" represents ability of the proposed global placement technique satisfying a good percentage of legalization rules for various cells."

"Global placement does not satisfy all the design and control-set rules."

"It shows how the placement results are improved gradually in terms of legalization and control-set rules along with congestion."

"After detailed placement, all the LUTs and FFs are placed as per the legalization and control-set rule."

"It can be inferred from the run time of the proposed algorithm that three phases, global placement, legalization, and detailed placement utilize roughly 55–60%, 20–25%, and 10–20% of total time."

"These heatmaps depict that how the proposed placer spread the cells over the placement regions to satisfy design rules."

Conclusion

"This paper presents an efficient wirelength-driven placement for UltraScale FPGAs."

"The proposed algorithm has a novel fixed block-based initial placement which directs the placement approach from the primary stage of the initial placement."

"Through three different phases, the proposed algorithm produces placement solutions for heterogeneous FPGAs with optimized wirelength."

Accelerating Computation of Steiner Trees on GPUs [6]

This is a machine-generated summary of:

Muniasamy, Rajesh Pandian; Nasre, Rupesh; Narayanaswamy, N. S.: Accelerating Computation of Steiner Trees on GPUs [6]

Published in: International Journal of Parallel Programming (2021)

Link to original: https://doi.org/10.1007/s10766-021-00723-0

If you want to cite the papers, please refer to the original.

For technical reasons we could not place the page where the original quote is coming from.

Abstract-Summary

"The Steiner Tree Problem (STP) is a well studied graph theoretic problem."

"One of the most popular algorithms is by Kou, Markowsky and Berman (KMB) which provides a 2-approximation to STP."

"A naïve implementation of the KMB algorithm is prohibitively slow for large graphs."

"Our goal in this work is to improve KMB algorithm's practical utility by parallelizing it on GPU and reduce its execution time on real-world graphs."

"This parallelization faces several challenges due to the inherent irregular nature of computation, as well as critical design decisions affecting the algorithm choice and optimizations."

"We overcome these challenges with interesting algorithmic observations and by exploiting parallelization within sub-steps, and develop the first GPU-based efficient approach to computing Steiner trees using KMB."

Introduction

"STP is stated as follows: A minimal-connected subgraph (that is, a tree) of G that contains all the terminals is called a Steiner tree T. The minimum weighted Steiner tree is called the optimum Steiner tree."

"The goal of the Steiner Tree Problem (STP) is to compute a subgraph of G which contains terminals L and the sum of the edge-weights in the subgraph is minimized."

"We tame these challenges by exploiting various invariants satisfied by the Steiner trees as well as by the KMB algorithm."

"Contributions This paper makes the following contributions: We develop the first GPU-based optimized implementation of the KMB algorithm to compute 2-approximate Steiner trees."

"We illustrate the challenges faced in the parallelization and show how to exploit various properties of Steiner trees as well as of the KMB algorithm."

Sequential KMB Algorithm

"It has three phases: (i) compute the shortest paths between all pairs of terminals, (ii) using the computed shortest distances, build another graph with edge-weights as the shortest distances and compute an MST on this new graph, and (iii) replace each MST edge in the new graph by the corresponding shortest path in the original graph and compute an MST in this graph, which is a Steiner tree of the original graph."

"The first phase computes the shortest path distances between every pair of terminals of G (for loop at Line 3)."

"Each SSSP populates two n-sized arrays: distance d and parent p. The d-array stores the shortest distance of each vertex from the source terminal, while p stores a parent in the SSSP tree (note that we say a parent rather than the parent, since there could be multiple shortest paths to a vertex; any one of them can be stored)."

Parallelizing the KMB Algorithm

"Algorithmic Choice KMB involves computing the shortest distance between every pair of terminals."

"We choose to use Bellman-Ford algorithm for SSSP computation, with the expectation that the parallelism benefits will outweigh the extra work done (which happens in practice)."

"Former research has also parallelized SSSP computation using variants of Bellman-Ford algorithm."

"An outer-parallel design allows us to use Dijkstra's algorithm, and retain the advantage of early SSSP halt (as discussed earlier)."

"Incorporating various design choices discussed so far, we implement the GPU version of the KMB algorithm using parallel subroutines for SSSP and MST."

"Discussion While there exist several studies optimizing parallelization of a single primitive computation (such as breadth-first search, shortest paths computation, page rank and triangle counting), our work involves building upon such graph primitives to design an efficient higher-level algorithm implementation (similar to computation of betweenness centrality)."

GPU KMB Optimizations

"In a pull-based approach, on the other hand, a vertex receives distance information from its neighbors and updates it locally as shown in Algorithm 5."

"Irrespective of push vs. pull approach, since the same vertex may be the source for one thread and target for another, there may be a read-write dependency."

"As an additional note, in both pull-based and push-based approaches, all the vertices are processed until the fixed-point is reached which is called as topology-driven approach [7]."

"In the pull-based approach, no atomic instruction is used while updating distances (Line 7 in Algorithm 5)."

"We optimize KMB's computation by (i) using a p-SSSP (ii) a data-driven approach, (iii) computation unrolling, (iv) storing the last state of the computation, and (v) modeling the computation using edges."

"When a vertex finds its distance updated, the corresponding thread propagates it to its neighbors immediately, without awaiting another kernel invocation."

Experimental Evaluation

"The execution time is affected by a complex interplay of several factors such as the overall graph size, number of terminals, degree distribution, diameter over terminals, maximum weight, etc But interestingly, these factors affect the algorithm's running time on CPU and GPU differently."

"Max-weight vs. performance To understand if maximum edge-weight of a graph affects the CPU or GPU performance, we compare the time taken by KMBCPU and KMBGPU-OPT on t1–t14 graphs with edge-weights and without edges-weight (i.e., unit edge weight)."

"The rest of the columns list deviation percentage of the resultant Steiner tree value for KMBCPU, KMBGPU-OPT, JEA and PACE with respect to the optimum value (abbreviated with single letters as C, G, J and P for brevity)."

"When the max-degree of the graph is very low (less than 5) or when the number of high-degree vertices is extremely large, we observe that the usefulness of shared memory optimization reduces."

Related Work

"Efficient and scalable implementations on GPUs have been considered for a wide variety of graph algorithms: Breadth First Search, Depth First Search, Minimum Spanning Tree, Single Source Shortest Path, All Pairs Shortest Path, Triangle Counting, Betweenness Centrality, to name a few."

“NP-Complete Problems GPU implementations of algorithms for NP-Complete problems is a relatively recent line of research.”

“The only approach to exactly solve NP-Complete problems is an exhaustive search of the solution-space in the worst-case (unless we have non-trivial understanding of NP-Complete problems), the prognosis for efficient parallel implementations is naturally weak.”

“Park and others [8] designed a parallel implementation of a classical 2-approximation algorithm, and discussed the challenges in parallelizing the KMB algorithm.”

“A recent survey [9] talks about the different approaches in the design of parallel algorithms for the STP.”

“Chow and others [10] had a GPU implementation for Rectilinear Steiner tree problem (which is a specialized version of STP).”

Conclusion and Future Work

“We have devised an efficient GPU implementation of the KMB algorithm by addressing challenges related to the graph representation, choice of parallelism, as well algorithm-specific GPU optimizations.”

“Our CPU implementation is an improvement over the OGDF’s KMB implementation, and it is on par with the PACE winner on the suite instances.”

“Based on its performance on a suite of graphs from competitive challenges as well as other real-world graphs, we conclude that the implementation provides a significant improvement of execution time over the sequential versions (JEA, PACE, and our KMBCPU).”

Faster Goal-Oriented Shortest Path Search for Bulk and Incremental Detailed Routing [11]

This is a machine-generated summary of:

Ahrens, Markus; Henke, Dorothee; Rabenstein, Stefan; Vygen, Jens: Faster goal-oriented shortest path search for bulk and incremental detailed routing [11]

Published in: Mathematical Programming (2023)

Link to original: https://doi.org/10.1007/s10107-023-01962-4

If you want to cite the papers, please refer to the original.

For technical reasons we could not place the page where the original quote is coming from.

Abstract-Summary

"We develop new algorithmic techniques for VLSI detailed routing."

"We improve the goal-oriented version of Dijkstra's algorithm to find shortest paths in huge incomplete grid graphs with edge costs depending on the direction and the layer, and possibly on rectangular regions."

"We combine the advantages of the two classical detailed routing approaches—global shortest path search and track assignment with local corrections—by treating input wires (such as the output of track assignment) as reservations that can be used at a discount by the respective net."

Introduction

"Each net is a set of pins and needs to be connected by a Steiner tree in the detailed routing graph."

"The subgraph is given by the global routing solution, excluding vertices and edges that would result in a conflict to previously routed wires."

"The second approach first considers the layers one after the other and assigns wires to routing tracks so that the most important detailed routing rules are satisfied, at least for most wires."

"This is achieved by finding a shortest path where reservations of the currently routed net can be used at a discount (so we reduce the cost of reserved edges by a fixed factor smaller than 1)."

"Treating the output of a track assignment as reservations (wherever it is legal) and then pursuing our new incremental routing approach can combine the advantages of the two classical bulk routing approaches, successive shortest paths and track assignment with local corrections."

Distances Without Preprocessing in the Simple Model

"There is a shortest path P between r and s in (G, c) that consists of at most one sequence of horizontal edges, at most one sequence of vertical edges, and hence at most three sequences of vias."

"If the walk is not a path, we can shortcut it to a path."

"Enumerate over all t rectangles that T is composed of."

"We compute the minimum cost of a path from r to s that (when traversed from r to s) consists of a path of vias, then a horizontal path, then a path of vias, then a vertical path, then a path of vias."

Logarithmic Query Time in the Simple Model

"The second ingredient is an algorithm for solving the planar point location problem, a well-studied problem in computational geometry."

"We initialize a data structure to solve the point location problem within that subdivision."

"If the query point is on the boundary of multiple polygons, it suffices to evaluate the minimizing function of any one of these polygons."

"If the query point is on the boundary of multiple tiles, it suffices to evaluate the minimizing function of any one of these tiles."

The General Model

"This is possible because all horizontal edges along some horizontal segment have the same cost, and hence, the corresponding labels can be represented by affine functions that can be stored and updated very efficiently."

"Of vertices we will store affine functions in a heap (priority queue), and the key of such a function will be the minimum relevant function value."

"We now propagate the labels from these vertices to the neighboring horizontal segments by computing at most six new affine functions: The algorithm stops when the heap is empty."

"We prove the assertion by induction on the order in which the affine functions are added (for all combinations of i, j, z, and x at once)."

"If f is added during the initialization phase, then there is a path with the desired property that only consists of horizontal edges."

Practical Aspects

"Shortest paths in the detailed routing graph often correspond to wirings that violate design rules."

"After a path search failed, BonnRoute may perform a backup path search which allows routing through existing wires at high cost (and then would remove (rip up) such wires and try to re-route them)."

"In chip design practice, there are two main scenarios where a detailed routing is not computed from scratch, using just a global routing as input, but in an incremental way, using an approximate detailed routing as input."

"The second scenario is when a step in between global and detailed routing is used, typically called track assignment, that maps the global wires to routing tracks in a way that obeys most—but not all—design rules."

"The traditional approach to incremental routing is to check for violations of design rules (e.g., wires of different nets overlapping) and to try to repair such violations locally, in a relatively small area around that violation."

A High-Quality Global Routing Algorithm Based on Hybrid Topology Optimization and Heuristic Search for Data Processing in MEC [12]

This is a machine-generated summary of:

Xu, Saijuan; Wei, Ling; Liu, Genggeng; Chen, Yeh-Cheng; Chen, Guolong: A high-quality global routing algorithm based on hybrid topology optimization and heuristic search for data processing in MEC [12]

Published in: The Journal of Supercomputing (2021)

Link to original: https://doi.org/10.1007/s11227-021-04147-y

If you want to cite the papers, please refer to the original.

For technical reasons we could not place the page where the original quote is coming from.

Abstract-Summary

"The existing work lacks the excellent optimization ability in wirelength, overflow, and runtime, or only considers some of the optimization objectives."

"We consider the wirelength, overflow, and runtime as the optimization objectives and propose a high-quality global routing algorithm of IC design in MEC environment, including the following effective strategies: (1) A hybrid topology optimization strategy combining Prim algorithm and divide-and-conquer method, (2) a heuristic search algorithm considering the congestion and the wirelength of nets."

"A heuristic search algorithm that considers both the congestion and the wirelength of nets is used to optimize the total wirelength in the R&R stage."

"In terms of the total overflow, the total wirelength and the runtime, the experimental results show that the proposed strategies have achieved effective optimization, which can better satisfy the demand of low delay of IC design for data processing in MEC environments of IoT."

Introduction

"RSMT is less flexible than RMST in the R&R stage, and too many Steiner points will lead to more concentrated and complex circuits in the network, which increases the congestion level and requires more time and cost to R&R. In addition, the adaptive multi-source multi-sink maze algorithm takes more into account the congestion problem and ignores the significant increase in wirelength when the net with overflow is rerouted."

"Experimental results demonstrate that this strategy can effectively reduce the number of overflows and the average running time. A heuristic search algorithm considering both the congestion and the wirelength of nets."

"In order to avoid the problem of increasing the wirelength in R&R stage, we propose a heuristic search algorithm that considers both congestion and wirelength, combining the advantages of the A* algorithm and the adaptive multi-source multi-sink maze routing algorithm, which can effectively reduce the congestion and wirelength."

Related Work

"The deployment strategy of MEC can realize the advantages of low delay and high bandwidth, and MEC can also obtain real-time wireless network information and more accurate location information to provide better services [13]."

"Liu and others [14] proposes X-architecture Steiner minimal tree algorithm based on multi-strategy optimization discrete differential evolution."

"Liu and others [15] proposes an effective two-stage X-routing Steiner minimum tree construction algorithm."

"Liu and others [16] proposes an efficient VLSI routing algorithm employing novel discrete PSO and multi-stage transformation to build X-architecture Steiner minimal tree and RSMT."

"Based on PSO, Liu and others [17] proposes a timing-driven octilinear steiner tree algorithm."

"Liu and others [18] proposes an obstacle-avoiding X-Architecture steiner minimal tree construction algorithm."

Problem Formulation

"The global routing usually transforms these rectangular cells into coarse-grained grid maps with capacity constraints, on which the nets are interconnected."

"For each net, the global routing needs to find a spanning tree that connects the vertices corresponding to all pins."

"For edge e, s(e) represents the maximum number of nets which it can pass through, and d(e) represents the actual amount of nets that pass through edge e. When the edge e has d(e) greater than s(e), the edge e is called to be an overflow routing edge and the net which it is located in has overflow."

Details of the Proposed Algorithm

"As the structure graph of global routing is usually dense, the FLUTE algorithm can be replaced by the Prim algorithm based on node selection in order to efficiently construct the minimum spanning tree."

"The algorithm outputs the node pointed to by min, which is the nearest neighbor node in the first quadrant of left, and moves left node down."

"For the RMST algorithm, in order to find the nearest neighbor node of each node in the vertex set in region 1, this can be done by properly transforming the point set definition and then using the above algorithm."

"The nearest neighbor nodes in each region of each point are calculated by eight calls to the method described in the previous section and then the RMST is constructed using the MST-based algorithm."

Experimental Results

"The analysis of the experimental results shows that the proposed strategy can further reduce the total overflow of the global routing results and reduce the running time to a certain extent, which shows that the strategy can effectively solve the problem of excessive congestion in FLUTE."

"In order to verify the effectiveness of the proposed heuristic search algorithm considering congestion and wirelength, this strategy is compared with the adaptive multi-source and multi-sink maze algorithm proposed by NTHU-Route 2.0."

"Since the proposed heuristic search algorithm considering both congestion and wirelength can simultaneously consider the influence of net on the congestion degree and the influence of wirelength in the iterative process of R&R, the purpose of reducing the wirelength of the net achieves."

"The experimental results fully demonstrate that our algorithm can effectively reduce the wirelength and run time."

Conclusion

"In order to optimize the wirelength, we propose a high-quality global routing algorithm of IC design."

"We propose a hybrid topology optimization strategy combined with Prim algorithm and divide-and-conquer method, which can effectively reduce the congestion of routing area."

"This paper proposes a heuristic search algorithm that considers both congestion and wirelength to effectively balance the impact of congestion and wirelength on routing results."

"The experimental results show that the above two strategies in this paper can effectively optimize the total overflow number, total wirelength, and average running time of the final routing results, so as to better satisfy the demand of low delay of IC design in MEC environments of IoT. The time delay is another important optimization index for VLSI routing, and we will consider the optimization of time delay in the layer assignment stage in the future work."

Genetic Algorithm Based Approach to Solve the Clustered Steiner Tree Problem [19]

This is a machine-generated summary of:

Do, Tuan Anh; Ban, Ha-Bang; Huynh, Thi Thanh Binh; Le, Minh Tu; Nguyen, Binh Long: Genetic algorithm based approach to solve the Clustered Steiner Tree Problem [19]

Published in: Evolutionary Intelligence (2023)

Link to original: https://doi.org/10.1007/s12065-023-00848-w

If you want to cite the papers, please refer to the original.

For technical reasons we could not place the page where the original quote is coming from.

Abstract-Summary

"A heuristic-based algorithm was proposed to solve the problem."

"This paper proposes a Genetic Algorithm called Shortest-Path Genetic Algorithm for solving the CluSteiner."

"The proposed algorithm can solve the problem in both Euclidean and non-Euclidean cases."

"Experiment results compared to existing works in the literature are analyzed in detail to prove the effectiveness of the proposed algorithm."

Introduction

"Steiner Tree Problem (STP) [20] is a well-known classical NP-Hard problem in optimization and graph theory."

"For their theoretical significance and real-world applications, many of the Steiner Tree problem variants have been developed and extensively studied, one of which is Clustered Steiner Tree Problem (CluSteiner) [21]."

"The CluSteiner belongs to the class of clustered graph problems, alongside others such as Clustered Traveling Salesman Problem (CluTSP) [22], Clustered Shortest-Path Tree Problem (CluSPT) [23], Minimum Routing Cost Clustered Tree Problem (CluMRCT) [24], etc These problems are distinguished in terms of their input graphs and objective functions."

"Motivated by the practicality of CluSteiner, we propose a first approximate method to tackle the CluSteiner problem, more specifically, using Genetic Algorithm (GA)."

"It has been utilized to solve various combinatorial optimization problems and has demonstrated effectiveness on some clustered graph problems [25, 26]."

"This paper proposes a novel Genetic Algorithm for the CluSteiner problem called Shortest-Path Genetic Algorithm (SPGA)."

Related Works

"Numerous studies have been carried out to address multiple variants of this problem; for example, the Euclidean Steiner Tree Problem is a popular NP-hard problem where the Steiner points lie on a Euclidean plane or the Rectilinear Steiner tree problem where the rectilinear distance is used instead of Euclidean distance."

"In [27], Chen and others proposed a heuristic algorithm to solve the Steiner Tree Problem in a multi-domain context."

"Since most versions of the Steiner Tree problems are NP-hard, popular approaches have been developing approximation algorithms, heuristics, or metaheuristics."

"The first one is the minimum inter-cluster Steiner tree problem (InterCluster), in which we want to find a cluster Steiner tree with minimum inter-cluster cost and ignore the local cost."

"The second is the minimum local-cost Steiner tree problem (LocalCluster) which asks for the minimum local cost among all clustered Steiner trees with minimum intercluster cost."

Preliminaries

"The remaining task is to solve the original Steiner Tree Problem on the newly contracted graph with the contracted vertices as required vertices."

"Any p-approximation algorithm of the Steiner Tree Problem can now be plugged in to solve on the reduced graph."

"The authors proposed to use an algorithm called SPH to solve the Steiner Tree Problem."

"They use SPH to build the Steiner tree that spans nodes belonging to different domains."

"The algorithm starts with a random node from the set of required vertices R and chooses it as the Steiner tree T. At each iteration, T grows by adding the shortest path from T to the nearest vertex of R that is not in T. This nearest vertex can be found using Dijkstra's algorithm with multiple starting points."

"When the graph is metric, the SPH algorithm is equivalent to the Prim's minimum spanning tree algorithm."

Proposed Algorithm

"Previous sections have introduced the Clustered Steiner Tree Problem."

"The main idea of this approach is to divide the CluSteiner into two subproblems: Finding the local trees of k clusters."

"This means a change in the order of clusters to construct a local tree would heavily affect the final result."

"The evaluation procedure of SPGA consists of two main steps corresponding to the two above-mentioned subproblems: Use SPH algorithm to construct the local tree of each cluster following the order determined by the priority permutation."

"Contract the graph by shrinking each cluster into a vertex, then use SPH algorithm to construct the inter-cluster links."

"Each evaluation finds the local Steiner tree of k clusters and the inter-cluster Steiner tree by using SPH."

"The order of clusters has no influence on the solution construction for each cluster's local tree."

Computational Results

"The improvement of the algorithm an over algorithm b is: To evaluate the efficiency of the proposed metaheuristic, we implement it on a benchmark dataset and compare it with the other algorithms."

"These algorithms can be compared with the same instances."

"The comparison with this algorithm will show how the evolution mechanism of GA improves solutions compared to just randomly exploring the search space."

"Each algorithm was simulated 30 times on each instance."

"To evaluate the performance in detail of the proposed algorithms, we perform two experiments as follows: Experiment 1: Analysis of the obtained results by the Non-parametric statistic."

"Non-parametric statistics are utilized to compare the efficacy of the algorithms BSPH, SPMST and the proposed algorithm SPGA."

"The results obtained by algorithms on types of instances are pointed out in Tables 14, 17, and 19."

"The time complexity by function evaluations: All algorithms run with the same number of fitness evaluations."

Discussion

"Good algorithms to solve this problem might not work for others."

"For NP-hard, three common approaches are usually used to solve the problem such as (1) exact algorithm, (2) approximation algorithm, and (3) heuristic or metaheuristic algorithm."

"The algorithm in [21] only solves the problem in the metric cases where all edge costs are nonnegative and satisfy the triangle inequality."

"To evaluate the efficiency of the proposed metaheuristic, we implement it on a benchmark dataset and compare it with the other heuristic and metaheuristic algorithms in the literature."

"There is only a (2+p)-approximation algorithm, where p is the best-known approximation ratio for the Minimum Steiner Tree Problem."

"Their algorithm only applies to the problem in the metric case."

"To fulfill the omission, we adjust the approximation ratio algorithm [21] with the Steiner Tree Solver being SPH to solve CluSteiner in metric and non-metric cases."

Conclusion

"The proposed algorithm introduces a new encoding method that decreases the chromosome dimensionality to the number of clusters and corresponding decoding."

"The proposed algorithm is applied to Euclidean and non-Euclidean cases well."

"Experiments and comparisons with several algorithms on numerous data sets were conducted to evaluate the proposal's efficiency."

OptiPlace: Optimized Placement Solution for Mixed-Size Designs [28]

This is a machine-generated summary of:

Datta, Prasun; Mukherjee, Shyamapada: OptiPlace: optimized placement solution for mixed-size designs [28]

Published in: Analog Integrated Circuits and Signal Processing (2021)

Link to original: https://doi.org/10.1007/s10470-021-01864-5

If you want to cite the papers, please refer to the original.

For technical reasons we could not place the page where the original quote is coming from.

Abstract-Summary

"A novel initial placement approach is introduced to achieve routability-aware global placement using new routability-aware cell clustering, cluster positioning, cluster movement, and pin offset based cell spreading methods."

"A routability-aware legalization has been implemented using new circle rolling and block sliding techniques."

"A routability-aware detailed placement is used to refine the placement in order to yield an optimized placement in terms of wirelength and routability for mixed-size designs."

"The detailed placement implemented by novel congestion hot spot detection, congestion contributing cells identification, congestion contributing cell movement techniques."

"A routing-congestion reduction method is used in detailed placement phase to refine the placement solution to achieve better routability."

"The solution quality of this proposed approach achieves improvised results than the contemporary in terms of HPWL, scaled wirelength, and routability."

Introduction

"Several previous approaches of probabilistic global routing algorithms [29–32] were devised for congestion estimation during placement."

"The analytic placement techniques [33–37] tries to change the objective function through additional optimization forces or costs to pull cells from the thick congested regions."

"For these two global placement benchmark suites, the leading-edge solution qualities of [38–42] shows improvisation over some widely accepted placers with congestion and routability awareness."

"The most acknowledged detailed placement techniques [43–47] modifies cell swapping objective to relieve congestion."

"Among various placement algorithms for VLSI circuits, wirelength-driven placers are not efficient enough to handle various design constraints."

"These routability-aware placement algorithms are more efficient than wirelength-driven placement to fulfil routing constraints during the routing phase."

"A routing overflow reduction is applied to refine the placement solution to achieve quality results with optimized wirelength and routing congestion."

Proposed Methodology

"Wirelength is expressed as half-perimeter wirelength (HPWL) as following: The objective of the proposed algorithm is to provide location of the cells/blocks in a predefined placement region of a circuit so that final routed wirelength is minimized and routing is possible with the available routing resources on the placement solution during the routing phase."

"In the cluster positioning step, clusters are placed at any location in the placement region based on the connections among clusters and fixed blocks."

"This increases overhead on the detailed placement phase and the total wirelength too, because a lot of cell swapping and movement are performed to reduce congestion and meet routability."

"Unlike this approach, the proposed algorithm finds out the congestion hotspots in the placement obtained after the second phase RaL. Now, instead of swapping a cell pair or moving cell instance to reduce congestion or overflow, a set of most congestion contributing cells are identified and moved to a less congested nearby area."

Experimental Results and Discussion

"The basis of this comparison is the global routing solutions achieved from NCTU-GR 2.0 [48] on the placement results produced by OptiPlace."

"It is observable that for all of benchmark circuits, OptiPlace reports better results than Ripple with respect to HPWL and routing congestion (RC)."

"The results in this tables report HPWL, RC, and CPU time of the placement solution for each benchmark circuit after the Initial Placement, R-Legalization, R-Detailed Placement without Routing Overflow Reduction, and R-Detailed Placement with Routing Overflow Reduction separately."

"It is clearly observable that the results obtained after "R-Detailed Placement without Routing Overflow Reduction" step of our proposed technique is better than Ripple 2.0."

"The placement solution produced by OptiPlace show that for all the benchmark circuits, the scaled wirelength is better than ICCAD 2012 winner, NTUplacer4h, POLAR and Ripple."

Conclusion

"The main advantage of the technique is that it controls routing congestion along with other performance measuring parameters like HPWL and Scaled wirelength."

"The results achieved by the proposed placer are better than the state-of-the-arts in terms of HPWL, scaled wirelength, and routability."

"The proposed OptiPlace achieves 1.02% and 3.53% improvement in terms of HPWL and routing congestion over one of the recent placer Ripple 2.0."

"In terms of scaled wirelength, OptiPlace shows far better performance than the recent placers for mixed-size designs."

An Efficient VLSI Circuit Partitioning Algorithm Based on Satin Bowerbird Optimization (SBO) [49]

This is a machine-generated summary of:

Pavithra Guru, R.; Vaithianathan, V.: An efficient VLSI circuit partitioning algorithm based on satin bowerbird optimization (SBO) [49]

Published in: Journal of Computational Electronics (2020)

Link to original: https://doi.org/10.1007/s10825-020-01491-9

If you want to cite the papers, please refer to the original.

For technical reasons we could not place the page where the original quote is coming from.

Abstract-Summary

"In partitioning, a circuit is recursively divided into several subcircuits so that each can be efficiently and independently designed with the main objective of reducing the cut-cost."

"A novel partitioning algorithm based on the concept of satin bowerbird optimization (SBO) is proposed herein."

"The bioinspired SBO algorithm chooses the initial partitions randomly then utilizes the concepts of fitness value evaluation and group migration to improve the cut-cost."

"The performance of the proposed algorithm is evaluated using parameters such as the cut-cost and time complexity based on simulations with ISCAS'85 benchmark circuits."

Introduction

"The partitioning of the net-list of the circuit is a principal step in the physical design of VLSI circuits."

"At this level, the specifications of the design are evaluated and fixed by considering factors such as operability, performance, physical size, and fabrication technology."

"The design level of the circuit design process is concerned with the use of transistors, logic gates, wiring, etc The physical design level comes after the circuit design, providing the circuit structure and transforming it into a layout description."

"The first level of the physical design process is the partitioning of the circuit into subcircuits."

"In this level, the precise location of the components of each subcircuit in every partition is assessed."

"During the design process in this level, the I/O pins of the components and the connections among them in each partition are considered."

Literature Survey

"Several of the algorithms currently applied for parallel processing of circuit partitioning are not robust and are thus not well matched for direct use in VLSI realization."

"Baruch et al. [50] compared the use of a genetic algorithm (GA) with simulated annealing (SA) for partitioning and observed that GA required a shorter run time to produce identical results."

"The performance of genetic algorithms in terms of complexity, time, and cost was observed to be excellent in comparison with other prevailing algorithms for circuit partitioning by Deep et al. [51]."

"Gupta et al. [52] applied the GA to solve circuit partitioning after observing that traditional optimization algorithms could not address the NP hardness problem."

"Lawler [53] presents some simple effective algorithms for optimal partitioning to minimize the delay of logic circuits."

"An evolutionary-based genetic algorithm for circuit partitioning that offers a substantial enhancement in terms of performance has also been proposed [54]."

The Formulation of the Circuit Partitioning Problem

"Circuit partitioning instances are commonly expressed as graph partitioning problems."

"For a given problem, any of these algorithms will always result in identical answers."

"Several well-known stochastic algorithms can be extensively applied to partitioning problems."

"Circuit partitioning is performed using specific algorithms."

"The circuit partitioning problem comprises the partition of a target circuit into nonintersecting circuits and subconnections with a minimum of one logic gate per subcircuit."

Algorithms

"An offspring population of partitions is generated in each generation using a crossover operation between the two parent partitions at a particular crossover site."

"The mutation operation is also applied by modifying the genetic structure of some chosen partitions in the parent population of each generation."

"To develop an effective GA, besides an appropriate solution architecture, it is essential that the original solution population be a varied representation of the space to be explored, so that genetic operators such as crossover and mutation are effective."

"The original population for the GA is initiated from random solutions."

"The crossover operator produces offspring solutions from two parent solutions from the original population."

"A different evolution policy is adopted herein, where the whole population of offspring substitutes the same number of inferior parent solutions to form the next generation."

The Analysis of the Performance of the GA, FA, PSO, ACO and SBO Algorithms

"The final cut-set value for the c432 benchmark circuit is 78, 74, 84, 78, and 72 when applying the PSO, ACO, GA, FA, and SBO algorithm, respectively."

"The final cut-set value for the c499 benchmark circuit is 182, 182, 152, 152, and 146 when applying the PSO, ACO, GA, FA, and SBO algorithm, respectively."

"The cut size increases remarkably when applying the SBO algorithm, while the time complexity is least for the SBO algorithm in comparison with ACO, PSO, FA, or GA for all the cases of node complexity."

The Analysis of the Performance of the Circuits

"These analyses are done by employing the cadence tool on the Linux platform, which can be extended to other standard benchmark circuits also."

"That, after applying the SBO algorithm, there is a reduction in the hyper-path delay."

"The same analysis can be applied for higher-complexity circuits such as C432 and C499, yielding similar results."

Conclusions

"Various bioinspired heuristic algorithms such as ant colony optimization, particle swam optimization, the genetic algorithm, and the firefly algorithm are used to perform the partitioning."

"Different performance parameters such as the minimum cut-cost, interconnections, and time complexity are analyzed using ISCAS'85 benchmark circuits such as C17, C432 and C499, in addition to investigation of the delay and area."

"A novel partitioning algorithm based on the SBO concept is also proposed and its performance parameters studied."

Packing and Legalization Free Boolean Satisfiability-Based Placement Algorithm for Heterogeneous FPGAs [55]

This is a machine-generated summary of:

Mukherjee, Shyamapada; Purkayastha, Sharbani: Packing and Legalization Free Boolean Satisfiability-based Placement Algorithm for Heterogeneous FPGAs [55]

Published in: Arabian Journal for Science and Engineering (2021)

Link to original: https://doi.org/10.1007/s13369-021-06176-4

If you want to cite the papers, please refer to the original.

For technical reasons we could not place the page where the original quote is coming from.

Abstract-Summary

"Placement of heterogeneous FPGAs has been a complex problem for large designs."

"Like other design problems, the placement problem of heterogeneous FPGAs has become very hard with the increasing number of the various components and their inter connections."

"A Boolean satisfiability (SAT)-based packing and legalization free placement algorithm called SatPhF has been introduced for Xilinx UltraScale FPGA architectures."

"The proposed technique reduces the effort given for packing and legalization in the state-of-the-art placement algorithms."

"The proposed approach takes care of congestion to produce an optimized routability-aware placement solutions in terms of wirelength and congestion."

"This technique performs a novel placement row selection method for a set of cells connected by a net using 2-SAT formulation which is able to optimize wirelength without use of any expensive analytical optimization framework."

"A new legalized location finding SAT approach has been introduced to find suitable locations for different cells without performing separate legalization step like other state-of-the-art placement algorithms."

"The proposed technique claims that most of the SAT instances created to formulate the placement problem is 2-SAT."

Introduction

"The placement problem for heterogeneous FPGAs has been formulated as SAT instances and solved to obtain routability-aware placement for them."

"The three main physical design steps of modern FPGAs are Packing of logic elements into BLEs, placement of packed components and other cells in different placement sites and routing of connecting nets through various routing channels."

"The proposed technique consists of three SAT-based steps: (1) Row Selection, (2) Cell Assignment, and (3) Placement Refinement."

"Our contributions are summarized below: A novel SAT-based placement algorithm for heterogeneous FPGAs has been proposed in this paper."

"A new 2-SAT instance-based formulation has been introduced to select a placement row for a set of cells belong to a net."

"A simple SAT-based approach for cell assignment to different possible legalized placement sites is presented in this paper."

Previous Work

"While RippleFPGA [56] consists of congestion estimation-based global placement with window-based legalization and routing resource-aware detailed placement to achieve routable placement results."

"In that paper, the authors propose an adaptive range-based algorithm which tries to improve the swap operations of simulated annealing algorithm to obtain better results in terms of critical path delay, speed and wirelength than simulated annealing algorithm. Reference [57] takes a different approach by partitioning the modules for placement."

"Machine learning-based approaches have been started to be used for placement and routing. Reference [58] is a machine learning-based algorithm to model the underlying relationship between the characteristics of circuits and the best CAD algorithms to obtain an optimized implementation on an FPGA."

"Some work on clock-driven FPGA architecture based on machine learning algorithm to solve the CAD problems is also been done."

"Pui and others [59] has targeted on clock-aware FPGA architecture that uses machine learning-based methods to predict routing congestion."

Target FPGA Architecture

"A heterogeneous FPGA consists of mainly four types of components: Configurable Logical Blocks (CLBs), Block Random Access Memories (BRAMs), Digital Signal Processing (DSPs), and Inputs/Outputs (I/Os)."

"In this architecture, a column provides logic resources for one specific type among CLB, RAM, DSP, and I/O. A placement site of one type may contain more than one cells in it."

"Most of the columns are for CLBs because a heterogeneous FPGA contains large number of LUTs and FFs."

"In the given architecture, each CLB consists of 16 LUTs and 16 FFs where 2 LUTs and 2 FFs together can form a single BLE."

"In columns 29, 65, 102, and 139, the red lines are denoting sites for DSPs."

"The green lines in various columns denote BRAM locations in the architecture."

Preliminaries

"The placement problem for any circuit is represented by a netlist which consists of set of nets of I/O pins in various cells."

"The objective of the placement problem is to find position values for the vectors X and Y that minimize the total wirelength and satisfy other placement constraints."

"Wirelength is measured by the half-perimeter wirelength (HPWL) metric which is expressed as following: Since, HPWL is not differential everywhere, quadratic placers approximate it by squared Euclidean distance between cells."

Multi Level SAT Formulation of Placement Problem

"After the selection of row for all the cells and associated nets, legal placement sites are determined for the cells assigned to a row and their corresponding columns based on their type."

"A similar kind of SAT instances are formulated for each net contains any fixed cell and are solved separately to find appropriate placement rows for them."

"Placement sites are identified in a placement row for the cells belong to a net which is assigned to the row in the previous Row Selection phase."

"In the proposed Cell Assignment technique, SAT formulation is performed for the connected LUTs to find their BEL location in a SLICE on the row under consideration as per the legalization rules."

"In the SAT formulation of phase II of the proposed approach, there is no restriction imposed on the number of instances placed in a placement site based on their type or legalization rules."

An Example

"The SAT instance is solved with a SAT solver to get a satisfiable assignment to the Boolean variables which provides a placement row for a net and the cells belong to the net under consideration."

"Boolean Formulation of Cell Assignment: Based on the type of the instances belong to a net and the row assigned to it in the previous step, certain Boolean variables are created."

"Boolean Formulation of Placement Refinement: Suppose, after cell assignment, some sites are congested by accommodating more cells than its capacity or due to the more demand in routing channels."

"Like the cell assignment, appropriate sites are identified in the adjacent placement rows above and/or below the row in which the congested site has been identified."

Results and Discussion

"The obtained results are little behind the state-of-the-art in terms of run time, it exhibits the ability to provide good placement results in terms of wirelength."

"Since, different SAT instances created during a placement phase of the technique can be solved parallelly, the total run time can be reduced by running the SAT instances parallelly."

"The proposed technique produces better results w.r.t the state-of-the-art RippleFPGA [56], UTPlaceF [60], GPlace 3 [61] in terms of wirelength."

"Our proposed placement method SatPhF proves it ability to produce complete placement solutions with lesser or at par wirelength results with the state-of-the-art."

"SAT instances are formulated for the nets in a circuit one at a time in phase I. The number of Boolean variable created in phase I for any net can be calculated as follows: where B denotes number of Boolean variables created and R is the number of placement rows."

Conclusion

"The proposed technique has used simple Boolean SAT formulation of a very complex physical design problem of heterogeneous FPGAs."

"This novel routability-aware SAT-based placement algorithm formulates SAT instances at different stages to obtain good placement results."

"The experimental results depict that the smart but simple SAT steps can produce at par results with the state-of-the-art techniques in terms of routability."

"In terms of wirelength, the proposed technique obtains better results than most of the state-of-the-art placement algorithms."

"The proposed technique formulates SAT instances which are Satisfiable mostly and requires less amount of time."

A Survey on Partitioning Models, Solution Algorithms and Algorithm Parallelization for Hardware/Software Co-Design [62]

This is a machine-generated summary of:

Hou, Neng; Yan, Xiaohu; He, Fazhi: A survey on partitioning models, solution algorithms and algorithm parallelization for hardware/software co-design [62]

Published in: Design Automation for Embedded Systems (2019)

Link to original: https://doi.org/10.1007/s10617-019-09220-7

If you want to cite the papers, please refer to the original.

For technical reasons we could not place the page where the original quote is coming from.

Abstract-Summary

"In electronic design automation, hardware/software co-design significantly reduces the time-to-market and improves the performance of embedded systems."

"With the increasing scale of applications and complexity of hardware architecture of embedded systems, hardware/software co-design is still a research hotspot."

"As hardware/software co-design is a wide topic, this paper focuses on major developments of three important aspects related to hardware/software partitioning, which has great effects on the performance of embedded systems."

"Existing parallel algorithms for hardware/software co-design are discussed in details."

Introduction

"Due to the increasing need of raising the potential quality and shortening the development time of electronic products, and the arrival of computer aided design (CAD) tools, HW/SW co-design had become an important topic since 1980s because of the advantages for concurrent design of hardware and software at the system level [63, 64]."

"Another survey paper [64] discussed one recent decade's development of HW/SW co-design, which included three periods: first step, maturation and mainstream."

"Most of previous survey papers discuss the whole HW/SW co-design workflow, which consists of system specification, task graph modeling, cost estimation, HW/SW partitioning, HW/SW interface design, system synthesis and evaluation."

"This paper focuses on one important aspect of co-design issues, the HW/SW partitioning."

"This paper carefully discusses the parallel methods for HW/SW partitioning algorithms, which are not seriously discussed in previous survey papers."

Partition Modeling

"The reconfigurable computing of FPGA is also treated as hardware component in HW/SW co-design process [65]."

"If DPR feature is not considered, it is a typical HW/SW partitioning, in which the task node number executed by hardware is constrained by chip area."

"There are many performance metrics (or objects) to evaluate the HW/SW partitioning result, such as energy consumption, area, heat dissipation, cost, etc Most of metrics are conflicting, hence it is a general way to select one or several critical metrics."

"According to the number of metrics, HW/SW partitioning can be treated as single-objective optimization or multi-objective optimization."

"Finding a reducible sub-graph is an effective way to improve the performance of HW/SW partitioning [66, 67]."

"Since existing HW/SW partitioning problems rely on the estimation of system performance and cost metrics which cannot be 100% accurate, several works [68, 69] emphasized the uncertainty of resources, configurations and metrics of embedded systems."

Algorithms for HW/SW Partitioning

"The goal of HW/SW partitioning is to find an optimal solution under given constraints."

"HW/SW partitioning is treated as Constraint Optimize Problem."

"Other domain-specific heuristics approaches for HW/SW partitioning include GCLP and MIBS [70], HOP and CHOP [71] and the heuristics proposed by [72]."

"A GCLP method for HW/SW bi-partitioning is firstly proposed in [70]."

"In their method, GCLP was utilized for HW/SW partitioning of one application."

"The process of domain-specific heuristics is closely tied to the particular HW/SW partitioning problem."

"The common meta-heuristics discussed by many literatures for HW/SW partitioning are genetic algorithm (GA) [73–76], simulated annealing (SA) [66, 77–80], and tabu search (TS) [77, 81–83]."

"Other meta-heuristics, such as ant colony algorithm (ACO) [84, 85], artificial bees [86], particle swarm optimization (PSO) [87, 88], artificial immune algorithm [89] were also proposed for HW/SW partitioning."

"Most HW/SW partitioning are treated as constraint optimizing problems, it is common to integrate domain-specific aspects into the framework of meta-heuristics to find a solution."

"The second category of general heuristics is to reduce HW/SW partitioning problems into another classical NP hard problems, such as graph partitioning [90, 91], or knapsack problem [81, 82, 92–97]."

Algorithm Parallelization

"Before the emergence of multi-core CPU/many-core GPU, the limitation of the availability and usefulness of parallel computing platforms make parallel methods for HW/SW partitioning scarce."

"Since cluster computer systems are utilized to parallelize meta-heuristics for HW/SW Partitioning, it should be more promising to devise parallel meta-heuristics on multi-core CPU/many-core GPU for the problem, allowing for the usefulness and availability of parallel platforms."

"To reduce the running time of computing communication cost of each particle, multi-core CPU/many-core GPU were utilized to parallelize this process, respectively."

"To reduce the running time and improve the solution quality, TS for HW/SW partitioning on GPU was researched in [98, 99]."

"It is seen that previous parallel methods for HW/SW partitioning only used multi-core CPU or many-core GPU."

"To reduce the running time, a hybrid parallel strategy was proposed on both multi-core CPU and many-core GPU."

Conclusion and Future Work

"This paper discusses partitioning models, solution algorithms and algorithm parallelization for HW/SW Co-design."

"After analyzing classical and new algorithms for HW/SW partitioning, it can be concluded that there exists no algorithm with absolute advantage."

"After discussing existing parallel algorithms for HW/SW Co-design in details, it can be concluded that Multi-core CPU/Many-core CPU can enhance the algorithm performance under limited computation resources."

"From the model aspect, there is a huge gap between the existing methods for HW/SW partitioning and real-world applications."

"The deep gap is that the traditional mathematical modeling approaches and/or traditional formal methods are limited by highly complex applications, high dimensional data and highly non-linear fitting, such as that in the problem of complicated HW/SW co-design."

"Not only the co-design model, but also the solution algorithm can benefit from the new generation of AI and data science [100–104]."

Compression and Speed-Up of Convolutional Neural Networks Through Dimensionality Reduction for Efficient Inference on Embedded Multiprocessor [105]

This is a machine-generated summary of:

Fernández Brillet, Lucas; Leclaire, Nicolas; Mancini, Stéphane; Nicolas, Marina; Cleyet-Merle, Sébastien; Henriques, Jean-Paul; Delnondedieu, Claude: Compression and Speed-up of Convolutional Neural Networks Through Dimensionality Reduction for Efficient Inference on Embedded Multiprocessor [105]

Published in: Journal of Signal Processing Systems (2021)

Link to original: https://doi.org/10.1007/s11265-020-01616-0

If you want to cite the papers, please refer to the original.

For technical reasons we could not place the page where the original quote is coming from.

Abstract-Summary

"To solve this, we introduce a new Principal Component Analysis (PCA) based compression, which relies on an optimal transformation (in the mean squared error sense) of the filters on each layer into a new representation space where convolutions are then applied."

"Compression is achieved by dimensioning this new representation space, with an arbitrarily controlled accuracy degradation of the new CNN."

"PCA compression is evaluated on multiple networks and datasets from the state of the art and applied to a binary face classification network."

"To show the versatility of the method and its usefulness to adapt a CNN to a hardware computing system, the compressed face classification network is implemented and evaluated on a custom embedded multiprocessor."

Introduction

"The computational complexity of CNN-based solutions makes this transition from traditional CV to CNN-based embedded solutions a difficult challenge, requiring both the adaptation of dedicated energy efficient hardware and the simplification of algorithms."

"We present a novel CNN compression and speed-up method based on Principal Component Analysis (PCA) [106], which is first benchmarked on multiple CNNs and datasets from the state of the art and then implemented on an embedded multi-processor for a real application such as face detection, showing the effectiveness and interest of the proposed method."

"Additionally to parameter and computational reduction, PCA compression shows to be an useful way to tune a CNN layer size according to hardware constraints, such as parallelization and vectorization rates."

"Experiments conducted on state-of-the art CNNs and popular data-bases show the effectiveness of the PCA compression."

Related Work: Model Compression

" [107–109], this quantization can be performed on the weights with low impact in network accuracy, allowing to reduce weight size by a factor of 32 with respect to a 32-bit implementation."

"In order to alleviate this issue, [108] proposes multiplying the outputs with a scale factor to recover the dynamic range (where the scale factor is the average of the absolute values of the weights in the filter), they keep earlier layers in the network at higher precision and also employ batch normalization in order to reduce dynamic range of activations."

"In the same line of work, [110] shows that it is also possible to transform a commonly learned CNN into the spectral domain on the fly in order to exploit the computational gain obtained by performing multiplication instead of convolution, specially for small filters, where FFTs are inefficient."

CNN Compression Through PCA

"The energy threshold for choosing the size of the PCA basis is computed by comparing the energy content of the reconstructed version of the filters and the energy content of the original filter matrix, that is: More concretely, direct application of the PCA algorithm allows to decompose a set of filters in any given layer into a hierarchical set of basic filters along with a set of coordinates."

"Since convolution is associative, for a given input I, we can write: This means that, instead of directly convolving the basis filters with the coordinates to regenerate the reconstructed filters $f_{reconstructed}$ and applying this later to the input, we can directly apply the basic filters f_{basis} to the input feature map I, and convolve the result by f_{coord} in the so-called recombination layer."

"Applying convolutions in the PCA basis, leads to a virtual intermediate feature map in the PCA space, which is then recombined in order to get back to the original feature map space."

Naïve Filter Reconstruction

"For the same energy thresholds, relative error is higher and compression rate is lower in the case of CIFAR-100 as compared to the same networks but applied for CIFAR-10 classification."

"An explanation to the rapid decay of accuracy of some of the tested networks is that PCA compression does not induce the same error at each layer for the same energy threshold and the impact of these errors on the network accuracy also depends on the layer."

"Along the curve 'All layers', the first point corresponds to the original network, the second point shows the effect of applying PCA solely to the first layer of the network (all other layers remain unchanged), the third point shows the effect of applying PCA to only the first layer and the second layer, etc… At last, the rightmost point of the curve shows the trade-off between compression rate and network accuracy when PCA is applied to all of the network's layers."

Trained Filter Reconstruction

"More concretely, the idea is to fine-tune the set of coordinates in each layer in order to improve the accuracy, the starting point being the values computed by PCA."

"Experiments show that retraining allows a severe improvement in accuracy for a given compression rate with respect to direct application of PCA."

"Task complexity (i.e., number of output classes) still seems to be an important factor, since the same network (ResNet-32) at the same energy threshold (60) yields a ×2 compression rate for a 2.07% accuracy drop in the case of CIFAR-10 as compared to a ×1.78 compression rate for a 3.97% drop for CIFAR-100 after retraining."

"The results presented in this section show that PCA is an effective method for selecting important parameters within a network, but an additional fine-tuning step is greatly beneficial and allows the network to recover most of the accuracy."

PCA for Filter Removal

"Another plausible idea is that the PCA basis filters already contain all the necessary information and therefore the introduced recombination layer could be advantageously replaced by the next layer of the network."

"Applying the method to all layers in the network leads to worse results than in the case where the recombination layer is kept."

"Experiments we completely remove the recombination layer (only keeping the basis filters) and randomly initialize the next layer, while the rest of the network's layers remain fixed."

"In order to generate these curves, PCA has been applied separately to each layer in the network."

"In the case of the first layer, the 128 filters are replaced by the 11 first basis filters, leading to a ×11.64 gain in the number of parameters, operations and feature map amount in both this layer and the following one, without any loss of accuracy."

PCA Combined with Other Compression Methods

"As opposed to PCA, pruning requires the indexing of zero values and therefore memory and computation gains are not directly proportional to the number of removed parameters."

"Quantization can also be applied to a network which has been compressed by PCA and pruned."

"The network is then retrained to reach 88.78% accuracy."

"Cumulatively, the network is compressed from 6.88Mbytes originally to around 370Kbytes (×19 memory gain for a 3.9% accuracy loss)."

"It can be seen, that while pruning and quantization are more robust to accuracy degradation, the combination of all techniques offers a multiplicative compression factor while globally preserving network accuracy."

Applicative Use Case: Face Detection

"The embedded memories are sized to 64KByte in order to store both the BFCN model (weights and parameters) and the largest feature map."

"As PCA compression allows to choose any size of basis filter it could be used to align the BFCN model with the architecture parameters."

"In order to align the BFCN network with the hardware architecture and further reduce the computing time would be to force the number of filters in each layer to be a multiple of 8, i.e., the number of PE and the size of the vector accelerator, allowing to skip some code required to check for corner cases."

"As result, for 1 PE, the number of cycles is now reduced by a factor ×2.63 from the baseline implementation and by ×2.45 for 8 PEs, resulting in a 20% improvement when compared to no alignment and negligible impact on network accuracy."

Conclusions & Future Work

"The proposed method is also suitable for hardware implementations as it doesn't require additional complex hardware blocks and preserves parallelism."

"We have applied this methodology to classification networks from the state of the art and also to a custom binary face classification network."

"A binary face classification network is implemented on a custom hardware multiprocessor and its performance is evaluated for multiple configurations."

"The versatility of the proposed compression method allowed to improve the performance both by reducing the intrinsic complexity and total number of operations, and also by tuning the network to compensate for hardware overheads."

Fast DSE of Reconfigurable Accelerator Systems via Ensemble Machine Learning [111]

This is a machine-generated summary of:

Lopes, Alba; Pereira, Monica: Fast DSE of reconfigurable accelerator systems via ensemble machine learning [111]

Published in: Analog Integrated Circuits and Signal Processing (2021)

Link to original: https://doi.org/10.1007/s10470-021-01885-0

If you want to cite the papers, please refer to the original.

For technical reasons we could not place the page where the original quote is coming from.

Abstract-Summary

"Answering when the combination of general purpose processors (GPPs) and RAs can provide the expected performance at the additional area and energy cost demands an extensive design space exploration (DSE)."

"Performing DSE through hardware synthesis is an extremely time-consuming and costly task."

"High-level simulations are a faster and simpler DSE method at the cost of accuracy loss."

"The use of high-level simulation does not allow simulating all design solutions and meeting time-to-market."

"Machine learning (ML) has become a promising solution to provide robustness to the DSE of large hardware designs by predicting aspects such as performance and energy."

"Ensemble learning is a promising solution since it can use multiple models and combine their predictions."

"We employ the use of ensemble methods to simplify and speed up the DSE of GPPs with RAs."

Introduction

"Machine learning (ML) based predictive models have emerged as a promising solution to reduce DSE time by predicting aspects such as performance, area, power and/or energy of possible hardware designs in the design space."

"The main contributions of this work are: A novel approach based on machine learning algorithms to perform DSE of CGRA coupled to GPPs by quickly estimating performance of applications running on previously untested architectural configurations; The use of regression ensembles to improve the prediction model accuracy and throughput even considering the diversity of embedded system applications."

"Comparison among four recurrently used regression algorithms to perform DSE of embedded system architectures with three ensemble models; A solution based on Gradient Boosting ensemble learning model which is able to achieve an error prediction rate below 2% for some benchmarks and a prediction throughput of more than 6000 predictions per second."

Related Works

"The authors employed linear regression and artificial neural networks to predict area and performance of a MPEG4 application running in the target platform."

"The study in [112] uses Random Forest as ML method to predict performance of homogeneous SoC platforms."

"Although the aforementioned studies have made advances in DSE of embedded systems using ML algorithms to predict metrics as performance and energy, most of them evaluate the ML algorithms in platforms with homogeneous components, as processors or accelerators [113–115]."

"In addition to evaluating the machine learning algorithms using only error as metric, we also include the throughput (predictions per second) as a factor to be considered in machine learning model selection when predicting performance of embedded system architectures."

"This work is the first one to perform an extensive investigation of machine learning regression models that includes different ensemble learning methods for DSE of heterogeneous hardware platforms."

ML-Based DSE Approach

"In next subsections, we will present the key concepts and challenges of GPP and CGRA design; the regression machine learning algorithms and the regressor ensemble methods to be used in our evaluation."

"These methods improve the predictive performance of a single model by training multiple models and combining their predictions."

"The key idea is to build a model that once trained on a set of points in the design space, can predict the behavior of different configurations."

"For each DSE goal, i.e., performance and/or energy, results from each benchmark running in each configuration must be saved. Model training and model evaluation The results obtained from previous step must be formatted as a training dataset."

"Each dataset is passed as input for machine learning algorithms, and each machine learning algorithm execution generates one predictive model."

Case Study

"Superscalar processor parameters we evaluated the processor's microparameters that influence the most the dynamic scheduling of instructions and the processor's ILP [116] as processor issue width, reorder buffer size, instruction queue size, load/store queue size and number of physical registers."

"In this stage, we used instructions per cycle (IPC) as metric of each benchmark execution in each processor issue width configuration."

"Considering the results obtained after executing the benchmarks with each physical integer registers configuration, the values chosen to perform DSE with this microarchitectural parameters was between 64 and 96, considering that lower values implied a IPC below 1 for the evaluated benchmarks and values above implied in increasing IPC below 0.1."

"Although we observed improvement in performance when increasing the number of levels for some benchmarks, we did not perform experiments with more than 60 levels."

Conclusions and Future Work

"We presented a machine learning approach for predicting performance of a heterogeneous hardware design composed of general purpose processor and reconfigurable accelerator."

"To cope with the complexity of this design, we performed an investigation of regression algorithms in an attempt to find a candidate model to be used in DSE flow of such systems."

"The experimental results indicate Gradient Boosting method as the primary candidate to be used as predictive model in our DSE flow."

"As future work, we intend to evaluate the robustness of ensemble algorithms when performing transfer learning to predict the performance of unseen benchmarks running at investigated platforms."

Clustering Without Replication in Combinatorial Circuits [117]

This is a machine-generated summary of:

Donovan, Zola; Gutin, Gregory; Mkrtchyan, Vahan; Subramani, K.: Clustering without replication in combinatorial circuits [117]

Published in: Journal of Combinatorial Optimization (2019)

Link to original: https://doi.org/10.1007/s10878-019-00394-1

If you want to cite the papers, please refer to the original.

For technical reasons we could not place the page where the original quote is coming from.

Abstract-Summary

"The modern integrated circuit is one of the most complex products engineered to date."

"Very large-scale integrated (VLSI) circuit design now involves massive design teams employing state-of-the-art computer-aided design (CAD) tools."

"One of the oldest, yet most important CAD problems for VLSI circuits is physical design automation, where one needs to compute the best physical layout of millions to billions of circuit components on a tiny silicon surface (Lim in Practical problems in VLSI physical design automation, Springer, Dordrecht [118])."

"The process of mapping an electronic design to a chip involves several physical design stages, one of which is clustering."

"Even for combinatorial circuits, there exists several models for the clustering problem."

"We consider the problem of clustering in combinatorial circuits for delay minimization, without permitting logic replication (CN)."

"The problem of clustering for delay minimization when logic replication is allowed (CA) has been well-studied and is known to be solvable in polynomial time [53, 119]."

Introduction

"We consider the problem of clustering without replication in combinatorial circuits for delay minimization (CN)."

"The delays of the interconnections are determined by the way the circuit is clustered."

"Intra-cluster delays are associated with the interconnections between gates in the same cluster."

"Inter-cluster delays are associated with the interconnections between gates in different clusters."

"The delay of the overall circuit, with respect to a specific clustering, is the maximum delay among all paths that connect an input to any output in the clustered-circuit."

"The problem of clustering combinatorial circuits for delay minimization when logic replication is allowed (CA) is well-studied [53, 119]."

"In CA, the goal is to find a clustering of a circuit that minimizes the delay of the overall circuit."

"When replication is allowed, circuit elements may be assigned to more than one cluster."

Statement of Problems

"We represent such circuits as directed acyclic graphs and formulate the main problem using graph-theoretic terminology."

"That we consider in this paper, each interconnection has one of the following types of delays: (1) an intra-cluster delay, d, when there is an interconnection between two gates in the same cluster, or (2) an inter-cluster delay, D, when there is an interconnection between two gates in different clusters."

"The delay of the overall circuit is the maximum delay among all source to sink paths in the circuit."

"A clustering algorithm tries to achieve one or both of the following goals, subject to one or more constraints: (1) The delay minimization through the circuit [119]. (2) The minimization of the total number of cut-edges [120]."

"We study CN under the delay model described as follows: 1."

"The delay of an interconnection between two gates within a single cluster is d. 3."

Related Work

"The algorithm for the general delay model proposed in Murgai et al. [121] does not always return an optimal solution, although they specify the conditions under which their algorithm returns an optimum."

"Rajaraman and Wong [119] consider CA under the more general delay model proposed in Murgai et al. [121], and discuss an algorithm that returns an optimal clustering in polynomial time."

"In Yang and Wong [122], they propose an efficient clustering algorithm that achieves an optimal solution under either the area constraint only or the pin constraint only."

"They show that the multi-level clustering problem for delay minimization with replication is NP-hard."

"They also propose an efficient heuristic for two-level clustering, providing a trade-off between area and delay by controlling the amount of replication."

"They also show how their algorithm applies to the problem of area-constrained min-cut partitioning with replication."

"They present an efficient heuristic which makes use of the clustering algorithm described in Rajaraman and Wong [119]."

[Section 4]

"We obtain the main results that deal with the computational complexity of CN."

"Our reductions imply that CN is inapproximable within a certain factor."

"Let w(s) and w(t) be the weights of vertices s and t, respectively."

"We obtain the following inapproximability result."

"Suppose that I is a "yes" instance of 3-BP 1-in-3SAT."

"I is a "yes" instance of 3-BP 1-in-3SAT."

[Section 5]

"A matching of a DAG G is perfect if any vertex of G is incident to an edge from the matching."

"Construct a longest path P in G and add odd edges to I. Construct a set S initially consisting of the edges of P. Note that I is a perfect matching in G[S], the subgraph of G induced by S. This property of I in G[S] is maintained."

"Consider every pair of edges not from I forming a directed path of length 2 and check whether the pair is on any longest path of G. If so, G has no dominating matching."

"If l is even, we have on the other hand, if l is odd, then since G has no dominating matching, we have hence case 2G has a dominating matching I. Then since any path of length l has an edge from I, we have the proof of the theorem is complete."

Conclusion

"We studied the problem of clustering without replication in combinatorial circuits for delay minimization (CN)."

"We are interested in the following open problems: 1."

Optimization of Thermal Aware Multilevel Routing for 3D IC [123]

This is a machine-generated summary of:

Sivakumar, P.; Pandiaraj, K.; JeyaPrakash, K.: Optimization of thermal aware multilevel routing for 3D IC [123]

Published in: Analog Integrated Circuits and Signal Processing (2019)

Link to original: https://doi.org/10.1007/s10470-019-01513-y

If you want to cite the papers, please refer to the original.

For technical reasons we could not place the page where the original quote is coming from.

Abstract-Summary

"This happens because of the increased wire length and routing path for connecting the components in chip."

"It is essential to reduce the wire length for the purpose of enhancing the circuit speed and minimize the power dissipation."

"There may be a chance of temperature rise due to the heat generated by the slacked layers which also needs to be reduced."

"Several traditional approaches have addressed the issues of temperature rise and layer assignment."

"A new stochastic based genetic algorithm is proposed in this work for reducing the thermal estimations."

"The length of wire can be reduced by determining the minimal path so as to connect the components."

"The superiority of this proposed approach is proved with reduced wire length and temperature estimations."

Introduction

"Connecting the device on 3DIC for various layers is performed by through-silicon via (TSV) 3D integration."

"This in turn enhance the concept of 3D IC's design."

"The 3D IC's design is capable of improving the system performance thereby decreasing the delays of interconnects and reducing the logic density."

"Whereas NTC schemes have an order-of-magnitude lower power ensuing in decreased thermal problems and the demand on power delivery, 3D ICs will be useful in enhancing both physical design and architecture levels performance."

"Three dimensional Integrated Circuits (3D-ICs) have the potential offerings over the heterogeneous integration that is essential for more than Moore (MtM) technology."

"During the routing stage, the TSVs are get placed by finding whitespace in the entire 3D IC, therefore the quality of a routing process purely depends on the available whitespace after the completion of TSV placement."

Related Works

"Their working procedure along with their merits and demerits are illustrated as follows. Reference [124] described a novel approach for placing the multiple dies on the interposer to optimize their locations for reducing integration overhead in the Interposer-Based 3D ICs."

"The performance of this work is comparatively better than other methods to acquire efficient results in liquid cooling networks of 3D ICs."

"Mirzaie and others [125] presented an optimization approach on the complete 3D architecture for improving the signal integrity and power efficiency counting through-silicon via (TSV), and I/O interface channel in a 65 nm CMOS technology."

"Chen and others [126] elucidated novel approach in 3-D integrated circuits, through silicon via (TSV) which is one of the critical method for providing vertical connection and some reliability issues are also occurred such as undercut and misalignment."

"The results of the presented approach offered better reputability compared to the other techniques in both 2D and 3D ICs."

Proposed Work

"The flow diagram is shown below and the following techniques are utilized for the effective performance of the TSV based 3D IC system with the use of genetic algorithm."

"For minimizing the thermal and the wire length of TSV based 3D IC approach, the neoteric technique is implemented."

"Several layer based set-up is performed to pick up the shortest routing path with the reduced thermal and wire length of TSV based 3D IC design."

"While evaluating the fitness, the thermal and wire length is estimated for the random placement of TSV."

"Estimate both thermal and wire length for the random placement of TSV at fitness evaluation select the TSV position by updating the chromosome positions."

"Initialization the process to create the initial population of algorithm Selection it is the process of determining the appropriate parents in order to give a child Crossover changing the chromosome of individuals from generations to generations."

Performance Analysis

"The wire length estimation and the temperature estimation for the proposed method is evaluated and compared with various benchmark datasets."

"The ISPD98 Circuit Benchmark Suite: The MCNC frequently presented and retained standard sets for practice by the Design Mechanization community."

"The second major path in the present public set of standards has just above 25,000 cells which is measured minor by today's criterions [127]."

"The proposed MS-GA is compared with the existing techniques."

Conclusion

"It is essential to reduce the wire length for the purpose of enhancing the circuit speed and minimize the power dissipation."

"A new stochastic based genetic algorithm is proposed in this work for reducing the thermal estimations."

"The wire length can be reduced by determining the minimal path to connect the components."

"The superiority of this proposed approach is proved with reduced wire length and temperature estimations."

References

1. Purkayastha S, Mukherjee S (2022) PHetDP: a placement algorithm for heterogeneous FPGAs with delayed packing. Circuits Syst Signal Process. https://doi.org/10.1007/s00034-022-02159-4
2. Brenner U, Rohe A (2003) An effective congestion-driven placement framework. IEEE Trans Comput Aided Des Integr Circ Syst 22(4):387–394. https://doi.org/10.1109/TCAD.2003.809662
3. He X et al (2013) Ripple 2.0: high quality routability-driven placement via global router integration. In: 50th ACM/EDAC/IEEE Design Automation Conference (DAC). ACM, pp 1–6. https://doi.org/10.1145/2463209.2488922
4. Alhyari A, Shamli A, Abuwaimer Z, Areibi S, Grewal G (2019) A deep learning framework to predict routability for FPGA circuit placement. In: 29th International Conference on Field Programmable Logic and Applications (FPL). IEEE, pp 334–341. https://doi.org/10.1109/FPL.2019.00060
5. Pui C-W, Chen G, Ma Y, Young EFY, Yu B (2017) Clock-aware ultrascale FPGA placement with machine learning routability prediction: (Invited paper). In: IEEE/ACM International Conference on Computer-Aided Design (ICCAD). IEEE, pp 929–936
6. Muniasamy RP, Nasre R, Narayanaswamy NS (2021) Accelerating computation of Steiner trees on GPUs. Int J Parallel Prog. https://doi.org/10.1007/s10766-021-00723-0
7. Nasre R, Burtscher M, Pingali K (2013) Data-driven versus topology-driven irregular computations on GPUS. In: 27th IEEE International Symposium on Parallel and Distributed Processing, IPDPS 2013, Cambridge, MA, USA. IEEE, pp 463–474. https://doi.org/10.1109/IPDPS.2013.28
8. Park J, Ro WW, Lee H, Park N (2008) Parallel algorithms for Steiner tree problem. In: Third International Conference on Convergence and Hybrid Information Technology, vol 1. IEEE, pp 453–455. https://doi.org/10.1109/ICCIT.2008.167
9. Bezensek M, Robic B (2014) A survey of parallel and distributed algorithms for the Steiner tree problem. Int J Parallel Prog 42(2):287–319. https://doi.org/10.1007/s10766-013-0243-z
10. Chow W, Li L, Young EFY, Sham C (2014) Obstacle-avoiding rectilinear Steiner tree construction in sequential and parallel approach. Integration 47(1):105–114. https://doi.org/10.1016/j.vlsi.2013.08.001
11. Ahrens M, Henke D, Rabenstein S, Vygen J (2023) Faster goal-oriented shortest path search for bulk and incremental detailed routing. Math Program. https://doi.org/10.1007/s10107-023-01962-4
12. Xu S, Wei L, Liu G, Chen Y-C, Chen G (2021) A high-quality global routing algorithm based on hybrid topology optimization and heuristic search for data processing in MEC. J Supercomput. https://doi.org/10.1007/s11227-021-04147-y
13. Liu X, Huang G, Zhao Q et al (2014) iMashup: a mashup-based framework for service composition. Sci China Inf Sci 54(1):1–20
14. Liu G, Yang L, Xu S et al (2021) X-architecture Steiner minimal tree algorithm based on multi-strategy optimization discrete differential evolution. PeerJ Comput Sci 7:e473
15. Liu G, Chen X, Zhou R et al (2021) Social learning discrete Particle Swarm Optimization based two-stage X-routing for IC design under Intelligent Edge Computing architecture. Appl Soft Comput 104(6):107215

16. Liu G, Zhu W, Xu S et al (2020) Efficient VLSI routing algorithm employing novel discrete PSO and multi-stage transformation. J Ambient Intell Humaniz Comput 15:155
17. Liu G, Guo W, Niu Y et al (2015) A PSO-based timing-driven octilinear Steiner tree algorithm for VLSI routing considering bend reduction. Soft Comput 19(5):1153–1169
18. Liu G, Huang X, Guo W et al (2015) Multilayer obstacle-avoiding X-architecture Steiner minimal tree construction based on particle swarm optimization. IEEE Trans Cybern 45(5):989–1002
19. Do TA, Ban H-B, Huynh TTB, Le MT, Nguyen B (2023) Long genetic algorithm based approach to solve the Clustered Steiner Tree Problem. Evol Intell. https://doi.org/10.1007/s12065-023-00848-w
20. Prömel HJ, Steger A (2002) The Steiner tree problem a tour through graphs, algorithms, and complexity. Springer
21. Wu BY, Lin CW (2015) On the clustered steiner tree problem. J Comb Optim 30:370–386. https://doi.org/10.1007/s10878-014-9772-7
22. Chisman JA (1975) The clustered traveling salesman problem. Comput Oper Res 2(2):115–119
23. D'Emidio M, Forlizzi L, Frigioni D, Leucci S, Proietti G (2016) On the clustered shortest-path tree problem. ICTCS 1720:263–268
24. Lin C-W, Wu BY (2017) On the minimum routing cost clustered tree problem. J Comb Optim 33(3):1106–1121
25. Menéndez HD, Barrero DF, Camacho D (2014) A genetic graph-based approach for partitional clustering. Int J Neural Syst 24(03):1430008
26. Ding C, Cheng Y, He M (2007) Two-level genetic algorithm for clustered traveling salesman problem with application in large-scale TSPS. Tsinghua Sci Technol 12(4):459–465
27. Chen L, Abdellatif S, Gayraud T, Berthou P (2017) A Steiner tree based approach for the efficient support of multipoint communications in a multi-domain context. In: 2017 IEEE symposium on computers and communications (ISCC). IEEE, pp 316–321. https://doi.org/10.1109/ISCC.2017.8024549
28. Datta P, Mukherjee S (2021) OptiPlace: optimized placement solution for mixed-size designs. Analog Integr Circ Sig Process. https://doi.org/10.1007/s10470-021-01864-5
29. Lou J, Krishnamoorthy S, Sheng HS (2001) Estimating routing congestion using probabilistic analysis. In: Proceedings of the ISPD. ACM, pp 112–117
30. Cheng C-LE (1994) Risa: accurate and efficient placement routability modeling. In: Proceedings of the 1994 IEEE/ACM international conference on computer-aided design, ser. ICCAD '94. IEEE Computer Society Press, pp 690–695
31. Hou W, Yu H, Hong X, Cai Y, Wu W, Gu J, Kao WH (2001) A new congestion-driven placement algorithm based on cell inflation. In: Proceedings of the 2001 Asia and South Pacific design automation conference, ser. ASP-DAC '01. ACM, pp 605–608
32. Yang X, Kastner R, Sarrafzadeh M, Sarrafzadeh M (2001) Congestion reduction during placement based on integer programming. In: Proceedings of the 2001 IEEE/ACM international conference on computeraided design, ser. ICCAD '01. IEEE Press, pp 573–576
33. Spindler P, Johannes FM (2007) Fast and accurate routing demand estimation for efficient routability-driven placement. In: 2007 design, automation test in Europe conference exhibition. IEEE, pp 1–6
34. Tsota K, Koh C-K, Balakrishnan V (2008) Guiding global placement with wire density. In: 2008 IEEE/ACM international conference on computer-aided design. IEEE, pp 212–217
35. Kahng AB, Wang Q (2005) Implementation and extensibility of an analytic placer. IEEE Trans Comput Aided Des Integr Circ Syst 24(5):734–747
36. Jiang Z-W, Su B-Y, Chang Y-W (2008) Routability-driven analytical placement by net overlapping removal for large-scale mixed-size designs. In: Proceedings of the DAC. ACM, pp 167–172
37. Chuang YL, Nam GJ, Alpert CJ, Chang YW, Roy J, Viswanathan N (2010) Design-hierarchy aware mixed-size placement for routability optimization. In: Proceedings of the ICCAD. ACM, pp 663–668

38. Lin T, Chu C (2014) Polar 2.0: an effective routability-driven placer. In: Proceedings of the DAC. ACM, pp 1–6
39. He X, Wang Y, Guo Y, Young EFY (2016) Ripple 2.0: improved movement of cells in routability-driven placement. ACM Trans Des Autom Electron Syst 22(1):101–1026
40. Hsu MK, Chen YF, Huang CC, Chou S, Lin TH, Chen TC, Chang YW (2014) Ntuplace4h: a novel routability-driven placement algorithm for hierarchical mixed-size circuit designs. IEEE Trans Comput Aided Des Integr Circ Syst 33(12):1914–1927
41. Cong J, Luo G, Tsota K, Xiao B (2013) Optimizing routability in large-scale mixed-size placement. In: Proceedings of the ASP-DAC. ACM, pp 441–446
42. Datta P, Mukherjee S (2019) Architecture-aware routability-driven placer for large-scale mixed-size designs. IET Circuits Dev Syst 13(8):1209–1220
43. Pan M, Chu C (2007) IPR: an integrated placement and routing algorithm. In: Proceedings of the DAC. ACM, pp 59–62
44. Kim M-C, Hu J, Lee D-J, Markov IL (2011) A simPLR method for routability-driven placement. In: Proceedings of the ICCAD. ACM, pp 67–73
45. He X, Huang T, Xiao L, Tian H, Young EFY (2013) Ripple: a robust and effective routability-driven placer. IEEE Trans Comput Aided Des Integr Circ Syst 32(10):1546–1556
46. Hsu M-K, Chou S, Lin T-H, Chang Y-W (2011) Routability-driven analytical placement for mixed-size circuit designs. In: Proceedings of the ICCAD. ACM, pp 80–84
47. Zhang Y, Chu C (2009) Crop: fast and effective congestion refinement of placement. In: 2009 IEEE/ACM international conference on computer-aided design—digest of technical papers. IEEE, pp 344–350
48. Liu WH, Kao WC, Li YL, Chao KY (2013) Nctu-gr 2.0: multithreaded collisionaware global routing with bounded-length maze routing. IEEE Trans Comput Aided Des Integr Circ Syst 32(5):709–722
49. Pavithra Guru R, Vaithianathan V (2020) An efficient VLSI circuit partitioning algorithm based on satin bowerbird optimization (SBO). J Comput Electron. https://doi.org/10.1007/s10825-020-01491-9
50. Baruch Z, Cret O, Pusztai K (1999) Genetic algorithm for circuit partitioning. In: Fifth International Conference on Engineering of Modern Electric Systems: Section Computer Science and Control Systems, pp 19–23
51. Deep A, Singh B, Singh A, Singh J (2009) A simple efficient circuit partitioning by genetic algorithm. Int J Comput Sci Network Secur 9(4):272–276
52. Gupta N, Garg D, Gupta S (2012) Genetic algorithms based partitioning of VLSI circuit systems. Int J Comput Appl 2:15–19
53. Lawler EL, Levitt KN, Turner J (1969) Module clustering to minimize delay in digital networks. IEEE Trans Comput C-18(1):47–57
54. Shahookar K (1995) Mazumder: genetic multiway partitioning. In: IEEE 8th International Conference on VLSI Design, New Delhi, India. IEEE, pp 365–369
55. Mukherjee S, Purkayastha S (2021) Packing and legalization free Boolean satisfiability-based placement algorithm for heterogeneous FPGAs. Arab J Sci Eng. https://doi.org/10.1007/s13369-021-06176-4
56. Chen G, Pui C-W, Chow W-K, Lam K-C, Kuang J, Young EFY, Yu B (2018) RippleFPGA: routability-driven simultaneous packing and placement for modern FPGAs. IEEE Trans Comput Aided Des Integr Circuits Syst 37(10):2022–2035. https://doi.org/10.1109/TCAD.2017.2778058
57. Mao F, Zhang W, He B, Lam SK (2017) Dynamic module partitioning for library based placement on heterogeneous FPGAs. In: 2017 IEEE 23rd International Conference on Embedded and Real-Time Computing Systems and Applications (RTCSA). IEEE, pp 1–6
58. Grе́wal G, Areibi S, Westrik M, Abuowaimer Z, Zhao B (2017) Automatic flow selection and quality-of-result estimation for FPGA placement. In: 2017 IEEE International Parallel and Distributed Processing Symposium Workshops (IPDPSW). IEEE, pp 115–123

59. Pui CW, Chen G, Ma Y, Young EF, Yu B (2017) Clock-aware ultrascale FPGA placement with machine learning routability prediction. In: 2017 IEEE/ACM International Conference on Computer-Aided Design (ICCAD). IEEE, pp 929–936
60. Li W, Dhar S, Pan DZ (2018) Utplacef: a routability-driven FPGA placer with physical and congestion aware packing. IEEE Trans Comput Aided Des Integr Circuits Syst 37(4):869–882
61. Abuowaimer Z, Maarouf D, Martin T, Foxcroft J, Gréwal G, Areibi S, Vannelli A (2018) GPlace 3.0: routability-driven analytic placer for UltraScale FPGA architectures. ACM Trans Design Automation Electr Syst 23(5):1–33. https://doi.org/10.1145/3233244
62. Hou N, Yan X, He F (2019) A survey on partitioning models, solution algorithms and algorithm parallelization for hardware/software co-design. Des Autom Embed Syst. https://doi.org/10.1007/s10617-019-09220-7
63. De Michell G, Gupta RK (1997) Hardware/software co-design. Proc IEEE 5(3):349–365
64. Wolf W (2003) A decade of hardware/software co-design. Computer 6(4):8–43
65. Cardoso J, Hubner M (eds) (2011) Reconfigurable computing: from FPGAs to hardware/software co-design. Springer, Berlin
66. Henkel J, Ernst R (2001) An approach to automated hardware/software partitioning using a flexible granularity that is driven by high-level estimation techniques. IEEE Trans Very Large Scale Integr 9(2):273–289
67. Jiang G, Wu J, Lam SK, Srikanthan T, Sun J (2015) Algorithmic aspects of graph reduction for hardware/software partitioning. J Supercomput 71(6):2251–2274
68. Jiang Y, Zhang H, Jiao X, Song X, Hung WN, Gu M, Sun J (2012) Uncertain model and algorithm for hardware/software partitioning. In: 2012 IEEE computer society annual symposium on VLSI. IEEE, pp 243–248
69. Wang R, Hung WN, Yang G, Song X (2016) Uncertainty model for configurable hardware/software and resource partitioning. IEEE Trans Comput 66(10):3217–3223
70. Kalavade A, Lee E (1997) The extended partitioning problem: hardware/software mapping, scheduling, and implementation-bin selection. Des Autom Embed Syst 2(2):125–163
71. Kalavade A, Subrahmanyam P (1998) Hardware/software partitioning for multi-function systems. IEEE Trans Comput Aid Des Int 17(9):819–837
72. Wolf W (1997) An architectural co-synthesis algorithm for distributed, embedded computing systems. IEEE Trans Very Large Scale Integr 5(2):218–229
73. Arat P, Juhasz S, Mann Z, Orban A, Papp D (2003) Hardware–software partitioning in embedded system design. In: 2003 IEEE international symposium on intelligent signal processing. IEEE, pp 197–202
74. Saha D, Basu A, Mitra RS (1997) Hardware software partitioning using genetic algorithm. In: Proceedings of 10th international conference on VLSI design. IEEE, pp 155–160
75. Srinivasan V, Radhakrishnan S, Vemuri R (1998) Hardware software partitioning with integrated hardware design space exploration. In: Proceedings of design, automation and test in Europe. IEEE, pp 28–35
76. Zou Y, Zhuang Z, Chen H (2004) HW–SW partitioning based on genetic algorithm. In: 2004 IEEE congress on evolutionary computation. IEEE, pp 628–633
77. Eles P, Peng Z, Kuchcinski K, Doboli A (1997) System level hardware/software partitioning based on simulated annealing and tabu search. Des Autom Embed Syst 2(1):5–32
78. Lopez-Vallejo M, Lopez JC (2003) On the hardware–software partitioning problem: system modeling and partitioning techniques. ACM Trans Des Autom Electron Syst 8(3):269–297
79. Banerjee S, Dutt N (2004) Efficient search space exploration for HW–SW partitioning. In: Proceedings of the 2nd IEEE/ACM/IFIP international conference on hardware/software codesign and system synthesis. ACM, pp 122–127
80. Garg K, Aung YL, Lam SK, Srikanthan T (2015) KnapSim-run-time efficient hardware–software partitioning technique for FPGAs. In: 28th IEEE international conference on system-on-chip. IEEE, pp 64–69
81. Wu JG, Sun QQ, Srikanthan T (2012) Algorithmic aspects for multiple-choice hardware/software partitioning. Comput Oper Res 39(12):3281–3292

82. Shi W, Wu J, Lam S, Srikanthan T (2016) Algorithms for bi-objective multiple-choice hardware/software partitioning. Comput Electr Eng 50:127–142
83. Jemai M, Dimassi S, Ouni B, Mtibaa A (2017) A metaheuristic based on tabu search for hardware/software partitioning. Turk J Electr Eng Comput Sci 25(2):901–912
84. Wang G, Gong W, Kastner R (2006) Application partitioning on programmable platforms using the ant colony optimization. J Embed Comput 2(1):119–136
85. Ferrandi F, Lanzi PL, Pilato C, Sciuto D, Tumeo A (2013) Ant colony optimization for mapping, scheduling and placing in reconfigurable systems. In: IEEE NASA/ESA conference on adaptive hardware and systems. IEEE, pp 47–54
86. Koudil M, Benatchba K, Tarabet A, Sahraoui EB (2007) Using artificial bees to solve partitioning and scheduling problems in co-design. Appl Math Comput 186(2):1710–1722
87. Abdelhalim M, Habib S (2011) An integrated high-level hardware/software partitioning methodology. Des Autom Embed Syst 15(1):19–50
88. Guoshuai L (2014) Improved hardware/software partitioning algorithm based on combination of PSO and TS. J Comput Inform Syst 10(14):5975–5985
89. Zhang Y, Luo W, Zhang Z, Li B, Wang X (2008) A hardware/software partitioning algorithm based on artificial immune principles. Appl Soft Comput 8(1):383–391
90. Arat P, Mann Z, Orban A (2005) Algorithmic aspects of hardware/software partitioning. ACM Trans Des Autom Electron Syst 10(1):136–156
91. Tahaee S, Jahangir A (2010) A polynomial algorithm for partitioning problems. ACM Trans Embed Comput Syst 9(4):34
92. Govil N, Shrestha R, Chowdhury SR (2017) PGMA: an algorithmic approach for multi-objective hardware software partitioning. Microprocess Microsyst 54:83–96
93. Wu J, Srikanthan T, Yan C (2008) Algorithmic aspects for area-efficient hardware/software partitioning. J Supercomput 79(4):1204–1215
94. Wu J, Srikanthan T, Lei T (2010) Efficient heuristic algorithms for path-based hardware/software partitioning. Math Comput Model 51(7):974–984
95. Wu JG, Srikanthan T, Chen G (2010) Algorithmic aspects of hardware/software partitioning: 1D search algorithms. IEEE Trans Comput 59(4):532–544
96. Wu JG, Wang P, Lam SK, Srikanthan T (2013) Efficient heuristic and tabu search for hardware/software partitioning. J Supercomput 66(1):118–134
97. Quan H, Zhang T, Liu Q, Guo J, Wang X, Hu R (2014) Comments on algorithmic aspects of hardware/software partitioning: 1D search algorithms. IEEE Trans Comput 63(4):1055–1056
98. Hou N, He F, Chen Y, Zhou Y (2016) An adaptive neighborhood taboo search on GPU for hardware/software co-design. In: IEEE 2016 international conference on computer supported cooperative work in design. IEEE, pp 239–244
99. Hou N, He F, Zhou Y, Ai H (2017) A GPU-based tabu search for very large hardware/software partitioning with limited resource usage. J Adv Mech Des Syst Manuf 11(5):JAMDSM0060
100. Chen Z, Yeo CK, Lee BS, Lau CT, Jin Y (2018) Evolutionary multi-objective optimization based ensemble autoencoders for image outlier detection. Neurocomputing 309:192–200
101. Li HR, He FZ, Yan XH (2019) IBEA-SVM: an indicator-based evolutionary algorithm based on pre-selection with classification guided by SVM. Appl Math J Chin Univ Ser B 34(1):1–26
102. Wang H, Jin Y, Sun C, Doherty J (2018) Offline data-driven evolutionary optimization using selective surrogate ensembles. IEEE Trans Cybern. https://doi.org/10.1109/TEVC.2018.2834881
103. Jin Y, Wang H, Chugh T, Guo D, Miettinen K (2018) Data-driven evolutionary optimization: an overview and case studies. IEEE Trans Cybern. https://doi.org/10.1109/TEVC.2018.2869001
104. Wang H, Jin Y (2018) A random forest-assisted evolutionary algorithm for data-driven constrained multiobjective combinatorial optimization of trauma systems. IEEE Trans Cybern. https://doi.org/10.1109/TCYB.2018.2869674

105. Fernández Brillet L, Leclaire N, Mancini S, Nicolas M, Cleyet-Merle S, Henriques J-P, Delnondedieu C (2021) Compression and speed-up of convolutional neural networks through dimensionality reduction for efficient inference on embedded multiprocessor. J Sign Process Syst. https://doi.org/10.1007/s11265-020-01616-0
106. Shlens J (2014) A tutorial on principal component analysis. arXiv: 1404.1100
107. Courbariaux M, Bengio Y, David JP (2015) Binaryconnect: training deep neural networks with binary weights during propagations. In: Advances in neural information processing systems. ACM, pp 3123–3131
108. Rastegari M, Vicente O, Joseph R, Ali F (2016) Xnor-net: imagenet classification using binary convolutional neural networks. In: European conference on computer vision. Springer, Cham, pp 525–542
109. Courbariaux M, Bengio Y (2016) Binarynet: training deep neural networks with weights and activations constrained to +1 or −1. In: Proceedings of the advances in neural information processing systems (NIPS). IEEE
110. Lavin A, Gray S (2016) Fast algorithms for convolutional neural networks. IEEE
111. Lopes A, Pereira M (2021) Fast DSE of reconfigurable accelerator systems via ensemble machine learning. Analog Integr Circ Sig Process. https://doi.org/10.1007/s10470-021-01885-0
112. Lin C, Du X, Jiang X, Wang D (2016) An efficient and effective performance estimation method for DSE. In: 2016 International symposium on VLSI design, automation and test (VLSI-DAT). IEEE
113. Malazgirt GA, Yurdakul A (2017) Prenaut: design space exploration for embedded symmetric multiprocessing with various on-chip architectures. J Syst Archit. https://doi.org/10.1016/j.sysarc.2016.07.004
114. Singh G, Gómez-Luna J, Mariani G, Oliveira GF, Corda S, Stuijk S, Mutlu O, Corporaal H (2019) Napel: near-memory computing application performance prediction via ensemble learning. In: Proceedings of the 56th annual design automation conference 2019. ACM, p 27
115. Sotiriou-Xanthopoulos E, Xydis S, Siozios K, Economakos G, Soudris D (2016) An integrated exploration and virtual platform framework for many-accelerator heterogeneous systems. ACM Trans Embed Comput Syst 15:1–26
116. Ponomarev D, Kucuk G, Ghose K (2001) Reducing power requirements of instruction scheduling through dynamic allocation of multiple datapath resources. In: Proceedings 34th ACM/IEEE international symposium on microarchitecture, MICRO-34. IEEE, pp 90–101
117. Donovan Z, Gutin G, Mkrtchyan V, Subramani K (2019) Clustering without replication in combinatorial circuits. J Comb Optim. https://doi.org/10.1007/s10878-019-00394-1
118. Lim SK (2008) Practical problems in VLSI physical design automation. Springer, Dordrecht
119. Rajaraman R, Wong DF (1993) Optimal clustering for delay minimization. In: 30th ACM/IEEE design automation conference. IEEE, pp 309–314
120. Hwang LJ, Gamal AE (1995) Min-cut replication in partitioned networks. IEEE Trans Comput Aided Des Integr Circuits Syst 14(1):96–106
121. Murgai R, Brayton RK, Sangiovanni-Vincentelli A (1991) On clustering for minimum delay/area. In: 1991 IEEE international conference on computer-aided design digest of technical papers. IEEE, pp 6–9
122. Yang HH, Wong DF (1997) Circuit clustering for delay minimization under area and pin constraints. IEEE Trans Comput Aided Des Integr Circuits Syst 16(9):976–986
123. Sivakumar P, Pandiaraj K, JeyaPrakash K (2019) Optimization of thermal aware multilevel routing for 3D IC. Analog Integr Circ Sig Process. https://doi.org/10.1007/s10470-019-01513-y
124. Osmolovskyi S et al (2018) Optimal die placement for interposer-based 3D ICs. In: Proceedings of the 23rd Asia and South Pacific design automation conference. ACM, pp 513–520

125. Mirzaie N et al (2017) A performance-aware I/O interface for 3D stacked memory systems. In: 2017 IEEE 8th annual ubiquitous computing, electronics and mobile communication conference (UEMCON). IEEE, pp 369–371
126. Chen Y-G et al (2015) Novel spare TSV deployment for 3-D ICs considering yield and timing constraints. IEEE Trans Comput Aided Des Integr Circ Syst 34:577–588
127. Alpert CJ (1998) The ISPD98 circuit benchmark suite. In: Proceedings of the 1998 International Symposium on Physical Design. ACM

Chapter 4
Power Minimization and Power Grid Synthesis

Introduction by the Editor

Modern world integrated circuits are tightly constrained by their power dissipation. Ask process node shrinks day by day we are working in smaller Micron levels. Now the integrated circuits are not only limited by the functionality, but also by the power they dissipate while functioning and in idle mode. This chapter majorly focuses on power minimization and power related synthesis which can help reduce power consumption in integrated circuit design. The techniques discussed in this chapter range from low voltage distribution and high impedance fault detection in connection with high level power Estimation techniques. Fine-grained power modelling of multicore processors, High level power estimation technique in embedded hardware, Fast macro model for substrate in mixed signal VLSI design, Low voltage distribution network technology identification and high impedance fault detection in Power Distribution grid systems are the techniques discussed in this chapter.

With scaling of technology process node, millions and millions of transistors are packed in the same square distance in the layout. We need effective and exact power and thermal modelling to replicate the actual silicon. These models should also capture the non-linear Performance versus power relationship. Power models use first degree modelling techniques to determine core's performance to power graphs. Linear modelling techniques are seldom useful for complex multi-core multi-power domain architectures. To achieve high accuracy, a feed forward neural network (FFNN) is used. NSGA-II is used to model optimization with multiple objectives. FFNN is trained in performance versus power data for various out of order processor architecture.

Power estimation modelling is a must for power optimizations. High level synthesis tools provide design implementation frameworks but have challenges when it comes to power consumption modelling. Statistical based methods, simulations, learning approaches and measurement methods are four categories of techniques

A. S. Shastri (ed.), *Optimization Methods in VLSI Design*,
https://doi.org/10.1007/978-981-95-2457-0_4

used for power consumption estimation. Learning approaches have better performance with moderate modelling efforts. In simulations, the lower the efforts, the higher the scope of error and vice versa. The error range was broad compared to learning approaches.

Substrate modelling greatly benefits simulation of substrate noise in mixed signal systems. This merger with SPICE simulation helps get a clear understanding of pre-layout and post layout substrate noise. Compared to finite element methods (FEM) and boundary element methods (BEM), Substrate modelling with SPICE simulations, based on finite sheet resistor slicing, incorporates the effect of supply rail bounce due to wire inductance. It infers real estimates of substrate noise effects with amazing compute efficiency.

Switch cluster deployment improves reliability and effectiveness of power systems. This automates power distribution grids as well. This deployment consists of choosing drivers of a self-healing system based on average interruption duration, number of consumers, compensation and the probabilities of current to exceed the pickup limit. Assigning communication links for stating automated breakers in power grid. Switch cluster is based on decision theory and linear programming. Similar multiple objective algorithms optimize, primarily, choice of links, balancing implementation costs and network reliability which enhances network topologies to allocate more resources in fault prone areas.

Harmony search algorithm (HSA) helps balance power consumption. Finding the power route is challenging while looking for that important node in WSN. Path operator calculus centrality (SPOCC) optimizes centrality problems in routing by using HAS and high centrality node from PSO. This technique gives optimized routing paths with greatly reduced power consumption. This hybrid algorithm. Helps get better improvement in terms of delay reduction and a good amount of residual energy than existing techniques.

Network topology is important for the operation of low voltage power distribution network. As power distribution grids do not have so good automation and network design blocks are spread out into multiple clusters, it's difficult to achieve this connectivity. Outer cabinet, Transformer, meter box and branch box are the four-level topology used to identify technique for distribution networks. This method is based on constrained least square and graph theory to find out the network structure using low voltage circuit breakers.

Power micro grids are used in different shapes and form as it has several benefits. Implementation for the same can create a lot of challenges like fault detection which could severely affect these grids. Support vector machines are used to detect high impedance fault (HIF). This reduces fault detection delays and operational challenges.

Machine Generated Summaries

Disclaimer: The summaries in this chapter were generated from Springer Nature publications using extractive AI auto-summarization: An extraction-based summarizer aims to identify the most important sentences of a text using an algorithm and uses those original sentences to create the auto-summary (unlike generative AI). As the constituted sentences are machine selected, they may not fully reflect the body of the work, so we strongly advise that the original content is read and cited. The auto generated summaries were curated by the editor to meet Springer Nature publication standards. To cite this content, please refer to the original papers.

Machine generated keywords: grid, modeling, distribution, distribution network, network, energy, estimation, network topology, noise, distribute, link, distribution system, system base, methodology, electrical

Fine-Grained Power Modeling of Multicore Processors Using FFNNs [1]

This is a machine-generated summary of:

Sagi, Mark; Vu Doan, Nguyen Anh; Fasfous, Nael; Wild, Thomas; Herkersdorf, Andreas: Fine-Grained Power Modeling of Multicore Processors Using FFNNs [1]

Published in: International Journal of Parallel Programming (2022)

Link to original: https://doi.org/10.1007/s10766-022-00730-9

If you want to cite the papers, please refer to the original.

For technical reasons we could not place the page where the original quote is coming from.

Abstract-Summary

"The state-of-the-art for deriving such power information is mainly based on predetermined power models which use linear modeling techniques to determine the core-performance/core-power relationship."

"With multicore processors becoming ever more complex, linear modeling techniques cannot capture all possible core-performance related power states anymore."

"To achieve a high estimation accuracy while minimizing run-time overhead, we propose a multi-objective-optimization of the neural architecture using NSGA-II with the FFNNs being trained on performance counter and power data from a complex-out-of-order processor architecture."

"We show that relative power estimation error for the highest accuracy FFNN decreases on average by 7.5% compared to a state-of-the-art linear power modeling approach and decreases by 5.5% compared to a multivariate polynomial regression model."

Introduction

"Previous works have shown that the relationship between performance counters and dynamic core power can also be nonlinear at core-level at least for low time-resolutions (1 s) [2]."

"The increasing complexity of modern core architectures integrating hundreds of millions of transistors per core and the importance of effective P&T management increasingly necessitate fine-grained high accuracy power models capturing nonlinear performance counter/power relationships."

"Investigate the use of FFNNs for high rate power estimations on core-level and the associated run-time inference overhead."

"With this paper we make the following contributions toward fine-grained run-time power estimation of multicores: We explore FFNNs for power estimation on core-level with an estimation rate of 10 kHz."

"We optimize the number of layers and the number of neurons per layer of the FFNNs with the objective of minimizing both power modeling error as well as run-time overhead."

Related Work

"Works that focus on high estimation rates usually rely on linear models to describe the relation between performance counters and dynamic power and can be found for Intel or AMD multicore processors in [3–5], for IBM multicore processors in [6] and for embedded ARM multicore processors in [7–9]."

"They determined that an FFNN architecture with two hidden layers—20 neurons in the first layer and 5 neurons in the second layer—to provide accuracy improvements compared to state-of-the-art linear power modeling techniques."

"Our work focuses on high rate power estimations with estimation rates of 10 kHz and provides a methodology to systematically generate FFNN architectures with both high accuracy and low run-time overhead."

"To our work, core-level power models with high estimation rates (10 kHz) were not explored."

"To previous work, this paper focuses on generating FFNN-based power models accounting for non-linear effects with fine-grained spatial (core-level) and time (10 kHz) resolution."

Feed-Forward Neural Networks for Power Modeling

"Single-objective Hyperparameter Solution Space The goal of our single-objective optimization methodology is to find neural architectures minimizing the power estimation error."

"Multi-objective Hyperparameter Solution Space The goal of our multi-objective optimization methodology is to find neural architectures minimizing both the power estimation error as well as the run-time overhead."

"The hidden neuron parametrizations are then used in the second tenfold cross validation step where we generate an FFNN for each training/validation fold combination, i.e., 10 FFNNs for each FFNN size (shallow, mid-sized and deep) and choose those FFNNs for testing which performed best on their corresponding validation data."

"Last step, the approach is similar to the single-objective methodology with FFNNs generated for the different validation folds and the ones with the lowest RMSE chosen for assessment on the hold out data with the only difference that there can be multiple neural architectures in the Pareto front and thus multiple final FFNNs."

Experimental Setup

"We trace common performance counters similar to other state-of-the-art work, e.g., [3], with a sampling rate of 10 kHz."

"Some works use a larger number of performance counters—in the range of 50–100—to estimate power due to the possibility of the estimation error decreasing with an increasing number of performance counter inputs."

"We use the power and energy information provided by McPAT in our evaluation which generates power information on package, core, and core component granularity with microsecond resolution."

"The simulated power values from McPAT can have inaccuracies which either favor/are neutral/disfavor our approach compared to the other state-of-the-art approaches."

Evaluation

"The computational and memory overhead of run-time inference of the FFNNs for producing a single power estimation is assessed by the necessary number of MAC operations and the memory needed to store the 32-bit neuron weights."

"Compared to the linear regression model, the shallow FFNN needs approximately 60 times more MAC operations and 40 times more memory while the multi-x FFNNs optimized for both accuracy and overhead need on average 21 times more MAC operations and 20 times more memory."

"The multi-x FFNNs and the shallow FFNN can be implemented for run-time power modeling on today's multicore processors."

"Of the different FFNN architectures, the multi-x architectures offer significantly higher inference rates and simultaneously lower compute logic and memory usage compared to the shallow, mid-sized, deep architectures."

"Compared to the polynomial model, the relative estimation error of the shallow FFNN power estimator still decreases by 5.5%."

Conclusion

"Our multi-objective optimization approach offers additional neural architectures with clear trade-offs between accuracy and run-time overhead for multicore power modeling."

"The highest accuracy FFNN from our previous work has in comparison to the newly generated FFNNs still a higher accuracy, however, the additional FFNNs from this work offer on average 60% lower overhead while still outperforming state-of-the-art linear and polynomial modeling approaches."

"We show that a micro-controller-based implementation would allow the FFNN inference/power estimation for all of our optimized FFNN architectures to be executed at 10 kHz on each core with a maximum area overhead of 0.25% for large out-of-order cores."

"The run-time overhead optimized FFNN architectures leave more headroom for additional, e.g., power and thermal management, tasks on the proposed micro-controller."

High-Level Power Estimation Techniques in Embedded Systems Hardware: An Overview [10]

This is a machine-generated summary of:

Richa, Majdi; Prévotet, Jean-Christophe; Dardaillon, Mickaël; Mroué, Mohamad; Samhat, Abed Ellatif: High-level power estimation techniques in embedded systems hardware: an overview [10]

Published in: The Journal of Supercomputing (2022)

Link to original: https://doi.org/10.1007/s11227-022-04798-5

If you want to cite the papers, please refer to the original.

For technical reasons we could not place the page where the original quote is coming from.

Abstract-Summary

"Early power consumption estimation at design time is crucial for power optimization."

"The aim of this paper is to present an overview of high-level power estimation techniques currently available along with a comprehensive comparison between different methodologies and their applications on estimated models."

"This paper provides designers, interested in power consumption modeling, with knowledge on how to select best power estimation techniques applied to designated target models."

Introduction

"Several commercial high-level synthesis (HLS) tools have reached maturity, providing a new method to implement FPGA/ASIC designs, or at least some parts of it but at the same time creating challenges in power consumption estimation [11]."

"Modeling FPGA power consumption is a necessity in two scenarios."

"Having accurate and efficient power models can help to rapidly explore design choices and perform high-level simulations."

"Widening the scope, high-level power estimation and modeling methodologies crossed beyond FPGA and ASIC limits, opening new horizons for microcontrollers/microprocessors-based systems with different but various techniques and architectures (soft-cores, hard-cores, customized instruction set, systems on a single chip, etc)."

"That said, power estimation and modeling at high level became a necessity."

"This paper provides an extensive overview on high-level power estimation and modeling techniques."

Estimation Techniques

"Based on the covered literature, four major power consumption estimation techniques were extracted: (1) the simulation-based methods, (2) the learning-based approaches, (3) the statistical-based methods and (4) the measurement-based alternatives."

"In [12], the high-level power estimation model for SoC is based on extended and optimized event signatures."

"The proposed method is achieved by propagating predictors between the connected neural models in order to estimate dynamic power consumption of individual components."

"In [13], the authors propose a high-level power estimation model based on input signals statistics defined as: the input signal probability, the average input signal density and the spatial correlation."

"With [14, 15] is a power estimation prediction technique based on a statistical model."

"We list some of this technique's target models based on related work: The power characterization methodology described in [16] relies upon a set of measurements where both static and dynamic power are extracted."

Approaches Classification

"In order to sort and classify the aforementioned techniques and models, a set of metrics had to be defined."

"The characterization reveals the distinctive nature of a given target model and highlights its special features/characteristics such as capacitance, switching activity, resources, etc Estimation effort: a qualitative metric represented by either little effort (+), moderate effort (++) and considerable effort (+++)."

"Modeling effort: similarly to the estimation effort, this qualitative metric is also represented by either (+), (++) or (+++)."

"The table's entries were sorted based on the number of occurrences in the literature for both the estimation techniques and their respective target models in descending order."

Discussions

"Based on the covered literature, the following is a comprehensive comparison between various estimation techniques applied to common target models: For the white/black-box IPs target models, the estimation error was found to be almost inverse proportional to the estimation effort targeting the simulation-based methods; the error range was higher and wider than the learning-based counterparts."

"Considerable estimation efforts in simulation-based methods definitely lead to satisfactory estimation % error, highly comparable to the learning-based alternatives, yet providing better results in some specific cases."

"As a global comparison, the learning-based techniques provide better performance in estimating power consumption at moderate efforts."

"For the component/circuit target models, a smaller and narrower range of estimation error was clearly detected with the learning-based methods."

"Little to moderate estimation efforts lead to estimation errors around 3.5% for the processor-based targets and 12.5% for IP targets."

Conclusion

"High-level power estimation techniques and target models were presented as the main concern and purpose of this paper."

"The purpose of that representation was to have a general and wide understanding of which estimation technique applied to which target model results in a better estimation accuracy while keeping an eye on required estimation efforts."

"The ultimate purpose of this work was to provide guidance for designers (dealing with power estimation and modeling methods) in terms of assessment, comparison and selection of the appropriate power estimation techniques applied to target models based on performance, accuracy and efforts."

A Fast Macro-Model for Substrate Coupling and Ground Bounce Induced Substrate Noise Simulation in Mixed Signal VLSI [17]

This is a machine-generated summary of:

Hasan, S.M. Rezaul; Shoaei, Omid: A Fast Macro-Model for Substrate Coupling and Ground Bounce Induced Substrate Noise Simulation in Mixed Signal VLSI [17]

Published in: Analog Integrated Circuits and Signal Processing (2003)

Link to original: https://doi.org/10.1023/a:1026261523831

If you want to cite the papers, please refer to the original.

For technical reasons we could not place the page where the original quote is coming from.

Abstract-Summary

"This paper discusses a novel substrate modeling technique for the simulation of substrate noise in mixed signal VLSI systems."

"This model yields to easy merger with the SPICE simulation netlist for the complete pre and post layout estimation of substrate noise effects in large mixed signal VLSI chips."

"Simulation results indicate a typical increase in Total Harmonic Distortion (THD) by atleast 6 dB due to the substrate noise effects, which corresponds to a performance loss by around 1-b precision."

"The substrate noise effects are found to be proportional to the oversampling ratio (i.e., the digital clocking rate with respect to the input signal) and the net number of logic transitions at each register transfer instance in the mixed signal chip."

Algorithms for Selecting and Interconnecting Switches to Automate Power Grids Considering Continuity Indexes and Reliability [18]

This is a machine-generated summary of:

Vaz, Rafael R. C.; Franco, Ricardo A. P.; Corrêa, Henrique P.; Vieira, Flávio H. T.; Araújo, Sérgio G.: Algorithms for Selecting and Interconnecting Switches to Automate Power Grids Considering Continuity Indexes and Reliability [18]

Published in: Journal of Control, Automation and Electrical Systems (2019)

Link to original: https://doi.org/10.1007/s40313-019-00522-6

If you want to cite the papers, please refer to the original.

For technical reasons we could not place the page where the original quote is coming from.

Abstract-Summary

"It is proposed to consider statistical analysis of current values using decision theory and binary linear programming to determine the priorities of switch clusters."

"The problem of allocating communication links for commanding automated breakers in a distribution power grid is also addressed."

"A novel heuristic is proposed for attributing reliability values to the links in terms of the subjacent power grid, inducing optimization convergence toward network topologies in which breakers at areas with higher fault indexes receive more communication resources."

Introduction

"These works do not perform neither the automation of the distribution network nor the determination of switch priority selection in self-healing systems."

"A methodology is proposed to decide which points and switches have the highest priority of choice to be used in the network automation according simultaneously to the SAIDI (System Average Interruption Duration Index), the number of consumers, the compensation and to the probabilities of the current to exceed the pickup limit."

"In resume, the overall proposed methodology for deployment switches to automate the power grid consists of effectuating a switching availability analysis using a power flow algorithm, discarding points in the network that do not permit the switching due to voltage conditions, determining priorities of the switch cluster and next applying an algorithm to optimize the fiber optics links allocation considering network reliability."

Methodology for Selecting Switch Clusters

"A study for determination of the clusters is conducted, observing which switching is possible if a fault occurs and considering the already established pickup current."

"Switch clusters are classified by the availability of switching regarding three metric types: maximum achieved current after load transference, maximum current at the time of greatest demand (8 a.m. to 7 p.m.) and probability of error type I (which will be explained later in this section)."

"The selection of switch clusters is determined by applying a binary linear programming (BLP) algorithm considering the System Average Interruption Duration Index (SAIDI), the compensation made to consumers due to problems in the network and the number of downstream consumers."

"This step will assign scores to each cluster, due to the statistical analysis, allowing to probabilistically determine which switching may occur, analyzing the post-switching current before it occurs."

Network Reliability Computation

"The associated communication network shall consist of n nodes (located at the breaker positions) and a certain number of fiber optics links between such nodes."

"The application of Yen's algorithm requires an auxiliary routine for obtaining the shortest path."

"A pseudocode of the proposed routine for calculating C is given in Algorithm 1."

"The optimization algorithm shall converge to solutions in which link redundancies are allocated mainly to such nodes."

"In the case study to be further presented, the considered automated breakers protected main feeders to which multiple smaller lines were connected."

[Section 4]

"Alongside reliability, total network cost will be taken as an objective function for optimization."

"This algorithm is attractive due to its output being a set of nondominated solutions, from which the decision maker can decide according to additional available information [19]."

"The NSGA-II algorithm also admits integration of penalty functions [19]."

Case Study: Results

"Considering the combination of SAIDI, number of consumers and total compensation, the cluster ranking order is: 8, 9, 10, 11 and 13, from the highest to the lowest priority of being chosen."

"The ranking order, from the highest choice priority to lowest, of the clusters in relation only to the probability of current exceeding the pickup current value is: 1, 3, 11, 4, 5, 6, 9, 7, 10, 8, or 13."

"Among the most important (priority) clusters in terms of SAIDI, number of consumers and total compensation, which are also less likely to exceed the pickup current, the result is: 1, 3, 4, 5 and 11."

"The results of the ranking of clusters show that the order of priority selection changes when considering the probability of exceeding the pickup current."

Conclusions

"A methodology was presented for the implementation of self-healing automation of distribution network with the objective of reducing System Average Interruption Duration Index per Consumer Unit in feeders."

"An algorithm was proposed for determining importance priorities of switch clusters in a self-healing system for an energy distribution network based on probabilistic analysis and binary linear programming."

"These indicators were used in order to classify the clusters and to show the possibility of performing the self-healing."

"A method was also proposed to optimize fiber optics link allocation for the automation of distribution breakers."

Analysis on Energy Consumption in Smart Grid WSN Using Path Operator Calculus Centrality Based HSA-PSO Algorithm [20]

This is a machine-generated summary of:

Hemalatha, R.; Prakash, R.; Sivapragash, C.: Analysis on energy consumption in smart grid WSN using path operator calculus centrality based HSA-PSO algorithm [20]

Published in: Soft Computing (2019)

Link to original: https://doi.org/10.1007/s00500-019-04580-5

If you want to cite the papers, please refer to the original.

For technical reasons we could not place the page where the original quote is coming from.

Abstract-Summary

"The consumption of energy is efficiently balanced using the high searching capability of harmony search algorithm (HSA)."

"Centrality in finding the route is considered as an utmost challenge in finding the important node in WSN."

"The SPOCC finds the main routing path using HSA, and high centrality node is estimated by particle swarm optimization (PSO) algorithm, thereby ensuring optimal routing with reduced energy consumption."

"The results show that the proposed hybrid algorithm obtains better improvement in terms of reduced delay and high residual energy than the existing algorithms."

Introduction

"It is further for optimizing the routing problems for finding the main path with high centrality nodes in WSN."

"In order to avoid the consumption of energy by the central nodes, the authors propose the shortest routing path using HSA-PSO algorithm."

"Faheem and Gungor [21] proposed EQRP proposed EQRP with bird mating optimization to improve the routing in Smart Grid Industry 4.0 WSN."

"Faheem and Gungora [22] proposed data capacity-aware channel assignment using the Fishbone routing algorithm in smart grid applications of WSN."

"The proposed study considers the centrality problem and addresses it using the proper selection of nodes for shortest routing paths in smart grid WSN environment."

"The SPOCC finds the main routing path using HSA and high centrality node is estimated by the PSO algorithm, thereby ensuring optimal routing with reduced energy consumption."

Shortest Path Operator Centrality Calculus

"The sensor nodes in smart grids are deployed randomly within the network range."

"The node with high centrality value is installed on the main path obtained from HAS-PSO routing algorithm, and it helps in establishing network connectivity."

"The SPOCC for a graph G with vertices V is estimated as follows: The optimal paths are estimated using HAS-PSO algorithm over each vertices pair (s, t), where the total energy of the node is assumed maximum and the bit error rate is assumed minimum."

"The optimal path for each SNs to reach the sink node is generally produced by BS, since it uses the SPOCC run estimation."

"The BS contains entire information about the SNs, i.e., it maintains centrality score of each node and another resource related information."

"The sensor nodes are deployed in a random manner and the size of the network presents the path operator centrality calculates the score."

HSA-PSO Algorithm for Finding the Shortest Path

"The PSO allows the swarm particles to move between the regions and updates the velocity and position after each iteration."

"For the given optimization problem, the rows in PHM is considered to be the random solution and hence the fitness function is calculated for each hybrid vector."

"Step 5 Updating particle position and velocity The optimal position of CH is determined by finding the best swarm particle with less cost function over n iterations."

"The particle best position and global best position is determined over each iteration, which is given by p_{best} and g_{best}."

"This is formulated as: where $V_{ij}(r)$ particle velocity at r, $X_{ij}(r)$ particle position at r, i ith particle, j dimension, r_1 and r_2 uniform random values between [0,1], w(r) inertia weight factor or linearly decreasing factor, c_1 and c_2 acceleration constants, p_{best} particle's local position, and g_{best} global best position."

Network Model

"The network model assumed in the present study uses a smart grid with 500 kV power grid station, which is the local power distribution area (LPDA)."

"The power distribution network has 1000 transmission grids used for evaluating the effectiveness of the proposed method."

"The distance of transmission between the individual grids is assumed to be 20 m in LPDA."

Results and Discussion

"Since this method avoids centrality within its range, the delay is minimal and the hybrid optimization algorithm provides the fastest routing process."

"The results show that the PDR of SPOCC-HSA-PSO is higher than the other methods."

"It is seen that the SPOCC-HSA-PSO has higher throughput range than the existing methods."

"It is seen from the results that the SPOCC-HSA-PSO has higher average residual energy."

"It is since the hybrid algorithm converges at a faster rate than the other algorithms and SPOCC enables faster routing process."

"The average increase in residual energy of the proposed system has increased to 14%, 21%, and 23% than multi-path QoS-aware routing algorithm, bird mating optimizer, and Fishbone routing algorithm."

Conclusions

"The smart grid infrastructure is considered unreliable and inefficient in terms of WSN communication, and hence a communication framework is required to improve the network reliability."

"With such indent, in this paper, the authors performed routing between the smart grids using SPOCC-HSA-PSO method by the real behavior of HSA and PSO individuals."

"The SPOCC operates using an operator calculus, which maintains the coverage, connectivity, and energy of the nodes in smart grid environments."

"The SPOCC is validated against individual WSN schemes and it is found that the proposed system is favored in smart grid applications."

Low-Voltage Distribution Network Topology Identification Based on Constrained Least Square and Graph Theory [23]

This is a machine-generated summary of:

Cui, Shijie; Zeng, Peng; Song, Chunhe; Wang, Zhongfeng; Li, Guangye: Low-voltage distribution network topology identification based on constrained least square and graph theory [23]

Published in: Soft Computing (2022)

Link to original: https://doi.org/10.1007/s00500-022-07151-3

If you want to cite the papers, please refer to the original.

For technical reasons we could not place the page where the original quote is coming from.

Abstract-Summary

"Network topology is essential for the safe operation of a low-voltage (LV) distribution network."

"We first propose a four-level topology (which includes transformer, outlet cabinet, branch box, meter box) automatic identification method for distribution networks."

"The proposed method is based on the constrained least square and graph theory to identify the network topology using LV intelligent circuit breaker energy measurements."

"The accuracy and effectiveness of the method are validated based on the practical application scenario of LV distribution networks and random simulation data."

Introduction

"We propose a new method to infer distribution network topology from LV intelligent circuit breaker (LVICB) measurements."

"Different from the previous methods that can only solve the connection relationship between users and transformers, the method proposed in this paper can identify the connection relationship between all electrical components and nodes in the LV distribution network as a whole, which provides a basis for fault location and line loss analysis."

"The contributions in this paper are threefold: (i) A four-level topology (transformer, outlet cabinet, branch box, meter box) automatic identification method for distribution networks is proposed."

"The linear regression analysis based on the law of energy conservation is used to realize the hierarchical division of network structure, which provides the basis for the accurate identification of network topology. (iii) The constrained least square algorithm and graph theory is used to identify topology from LVICB measurements."

Problem Description and Mathematical Expression

"The commonly used topological structure representation methods include incidence matrix and adjacency matrix [24]."

"The adjacency matrix represents the connection between nodes, which impressively avoids the problem of unknown branch conditions."

"The adjacency matrix does not consider the direction of power flow, so it can only represent whether there is a branch connection between nodes, and the tide direction on the branch cannot be determined."

"If the transformer, branch switch node, and incoming switch node of the consumer meter box on an LV distribution network are regarded as vertices, and the distribution feeder is regarded as edges, then the topological structure of the distribution network can be expressed by an adjacency matrix."

"In the steady state, the distribution network topology belongs to the acyclic directed graph, and its adjacency matrix is unique."

Proposed Approach for Topology Identification

"From the known measured data, the steady-state network topology of the LV distribution network is deduced, which is a typical data-driven method."

"It is assumed that all the network nodes can be observed without electricity stealing, and the network topology remains unchanged in the process of data acquisition."

"According to the conservation of energy, the nodes of layer 1 with known location are taken as the parent nodes, and the nodes of layer 2 adjacent to layer 1 are calculated."

"The connection relation between all nodes can be deduced and the topology identification can be realized."

"The solution result is rounded to obtain the connection relation between node 1 and other nodes."

"The connection relation between two adjacent nodes can be obtained by the least square method with constraints."

Experimental Results

"With the increase of the number of samples, the CLS algorithm quickly improves the success rate to 100%, and its overall performance is better than PCA."

"That the calculation time of the PCA algorithm increases significantly when the number of samples is more than 100 because its algorithm complexity is $O(M^2N)$, which is more suitable for the case of a large number of LVICB nodes and a small number of measurement samples."

"The calculation time of CLS increases linearly with the increase of the sample number, because its algorithm complexity is $O(N^2M)$."

"In general, when $N = 19$ and $M = 300$, the calculation time of CLS is only 0.054 s. In most cases, the number of circuit breakers in a distribution station area is about 50(Xiao and others 19), but the number of samples should be more than 300 to achieve better identification accuracy."

Conclusion

"This paper proposes a topology identification method for LV distribution network based on the CLS method and graph theory."

"The method is based on the existing hardware (LVICB), and needs no additional hardware equipment, which could reduce the cost and is implemented easily in practice. The proposed algorithm is a data-driven method with faster speed and better accuracy."

"Besides, the algorithm has low time complexity and can be easily deployed on the edge side. The proposed method provides a basis for the intelligent upgrading of the distribution station area and has practical application potentials."

High-Impedance Fault Detection in Power Distribution Grid Systems Based on Support Vector Machine Approach [25]

This is a machine-generated summary of:

Ahmadi, Ali; Aghajari, Ebrahim; Zangeneh, Mehdi: High-impedance fault detection in power distribution grid systems based on support vector machine approach [25]

Published in: Electrical Engineering (2022)

Link to original: https://doi.org/10.1007/s00202-022-01544-1

If you want to cite the papers, please refer to the original.

For technical reasons we could not place the page where the original quote is coming from.

Abstract-Summary

"Its implementation, however, makes significant theoretical and practical challenges, such as fault detection that could make crucial damage to the utility grid and microgrids."

"This paper, accordingly, developed a novel, fast and accurate method based on Support Vector Machines (SVM) approach for High Impedance Fault (HIF) detection."

"The proposed method applied to a typical distributed generation system for detecting single line, double line and triple line HIFs."

"This method will boost the development of renewable energy usage by reducing fault detection delays and operational risks."

Introduction

"Another method for detecting faults in the power systems is WT method, which has attracted many researchers in the field."

"With the development of high-rise buildings, the same method uses to detect the arc fault in the indoor distribution systems [26]."

"This paper developed an approach for HIFs detection in power grid transmission lines based on the SVM method."

"The ability to train in SVM method allows the fault detection algorithms in relays to be always updated; when necessary, proper training data can introduce new fault to the relays."

"The main contributions of this study can be organized as follow: (1) the designed protection scheme could successfully detect symmetric and asymmetric ground faults and separate them from usual normal transient states of the network grid. (2) The proposed SVM method, in addition to having strong theoretical support is also highly simple in performance and implementation."

Support Vector Machine Method

"The classification by SVM mainly encompasses data training and testing that is comprised of many cases."

"In the training set, any case includes two features and one target value named class label."

"The purpose of this procedure is to produce a system that can successfully determine the label of new data or test case data that only includes features."

"The majority of this method's algorithms, though, can divide data between just two classes hence make it a two-class problem [27–29]."

"Z determines the misclassification's amount in non-separable cases."

"The distance between misclassification and separating hyperplane is defined by $(-\zeta/||w||)$ [30]."

"The input data must first be mapped from the n-dimensional space to a higher dimension space, then one more function is applied to map all the data to the decision space [28]."

Power Distribution Grid System

"The line between buses A and B is regarded as a protected line and the fault detection system is on the bus A side."

"It assumed the faults occurred between buses A and B in the length of 50 KM."

"It is notable that the fault detection system detects the faults by processing three-phase currents."

"The proposed SVM method requires training data to detect and react correctly in fault situations."

Simulation Results and Discussion

"Recognized by the Wavelet and Park approach at 0.1028 s. Also, this fault was detected by ANN and SVM methods with a time difference of around 1 ms in 0.0063 and 0.0076 s, respectively."

"The SVM, ANN and wavelet and park methods detected this HIF in 0.06128, 0.06174 and 0.109 s, respectively."

"The second level of detail of the Wavelet method detects the fault very well, and by selecting 0.5 for the threshold value, the time required for detection will be 19 ms."

"The advantages of machine learning approaches and the recently mentioned superiorities of the proposed method introduce the SVM technique as a safe, fast and reliable choice for fault detection in distribution power systems."

Conclusion

"This paper presents a novel protection technique for the power distribution grid systems based on the SVM approach."

"The performance of the proposed protection scheme has been investigated by simulating a distributed power grid and applying different types of HIFs to this system using Matlab software."

"Three types of faults including triple-line, double-line and single-line to the ground considered in this research that the proposed method could successfully detect them and make an appropriate response in a reasonable time."

"It is expected that this effective technique offers more reliability and decreases the operational risk factor in distribution power systems, especially microgrids that include renewable energy power generation units."

References

1. Sagi M, Vu Doan NA, Fasfous N, Wild T, Herkersdorf A (2022) Fine-grained power modeling of multicore processors using FFNNs. Int J Parallel Program. https://doi.org/10.1007/s10766-022-00730-9
2. McCullough JC, Agarwal Y, Chandrashekar J, Kuppuswamy S, Snoeren AC, Gupta RK, Diego UCS, Labs I (2011) Evaluating the effectiveness of model-based power characterization. USENIX ATC
3. Bertran R, Gonzelez M, Martorell X, Navarro N, Ayguade E (2013) A systematic methodology to generate decomposable and responsive power models for CMPs. IEEE Trans Comput, 62(7), pp. 1289–1302. 14 pp
4. Bircher WL, John LK (2012) Complete system power estimation using processor performance events. IEEE Trans Comput, 61(4), pp 563–577. 15 pp
5. Chadha M, Ilsche T, Bielert M, Nagel WE (2017) A statistical approach to power estimation for x86 processors. In: Proceedings of the 2017 IEEE 31st international parallel and distributed processing symposium workshops, IPDPSW 2017. IEEE
6. Huang W, Lefurgy C, Kuk W, Buyuktosunoglu A, Floyd M, Rajamani K, Allen-Ware M, Brock B (2012) Accurate fine-grained processor power proxies. In: IEEE/ACM MICRO. IEEE
7. Rethinagiri SK, Palomar O, Ben Atitallah R, Niar S, Unsal O, Kestelman AC (2014) System-level power estimation tool for embedded processor based platforms. In: ACM RAPIDO. ACM
8. Su B, Gu J, Shen L, Huang W, Greathouse JL, Wang Z (2014) Ppep: online performance, power, and energy prediction framework and DVFs space exploration. In: IEEE/ACM MICRO. IEEE
9. Walker MJ, Diestelhorst S, Hansson A, Das AK, Yang S, Al-Hashimi BM, Merrett GV (2017) Accurate and stable run-time power modeling for mobile and embedded CPUs. IEEE TCAD
10. Richa M, Prévotet J-C, Dardaillon M, Mroué M, Samhat AE (2022) High-level power estimation techniques in embedded systems hardware: an overview. J Supercomputing. https://doi.org/10.1007/s11227-022-04798-5
11. Marjanovic J (2018) Low vs. high level programming for FPGA. In: Proceedings of the 7th International Beam Instrumentation Conference, IBIC 2018. IEEE, pp 527–533. https://doi.org/10.18429/JACoW-IBIC2018-thoa01
12. Wang L, Wang X, Wang T, Yang Q (2012) High-level power estimation model for SOC with FPGA prototyping. In: Proceedings—4th International Conference on Computational Intelligence and Communication Networks, CICN 2012. IEEE, pp 491–495. https://doi.org/10.1109/CICN.2012.124
13. Durrani YA, Abril A, Riesgo T (2007) Efficient power macromodeling technique for IP-based digital system. In: Proceedings—IEEE International Symposium on Circuits and Systems. IEEE, pp 1145–1148. https://doi.org/10.1109/ISCAS.2007.378252
14. Irfan M, Masud S, Pasha MA (2018) Development of a high level power estimation framework for multicore processors. In: Proceedings of 2018 2nd IEEE Advanced Information Management, Communicates, Electronic and Automation Control Conference, IMCEC 2018. IEEE. https://doi.org/10.1109/IMCEC.2018.8469473
15. Paniego JM, Libutti L, Puig MP, Chichizola F, De Giusti L, Naiouf M, De Giusti A (2020) Unified power modeling design for various raspberry pi generations analyzing different statistical methods. Commun Comput Inform Sci. https://doi.org/10.1007/978-3-030-48325-8-4
16. El-dib D, El-dib DA, Alkabani Y, El-moursy M, Shawkey HH (2016) Automated FPGA power characterization methodology. Int J Elect Comput Sci IJECS-IJENS. 16(2):25–32
17. Hasan SMR, Shoaei O (2003) A fast macro-model for substrate coupling and ground bounce induced substrate noise simulation in mixed signal VLSI. Analog Integr Circ Sig Process. https://doi.org/10.1023/a:1026261523831
18. Vaz RRC, Franco RAP, Corrêa HP, Vieira FHT, Araújo SG (2019) Algorithms for selecting and interconnecting switches to automate power grids considering continuity indexes and reliability. J Control Automat Elect Syst. https://doi.org/10.1007/s40313-019-00522-6

19. Deb K, Pratap A, Agarwal S, Meyarivan T (2002) A fast and elitist multiobjective genetic algorithm: NSGA-II. IEEE Trans Evolut Comput 6(2):182–197. https://doi.org/10.1109/4235.996017
20. Hemalatha R, Prakash R, Sivapragash C (2019) Analysis on energy consumption in smart grid WSN using path operator calculus centrality based HSA-PSO algorithm. Soft Computing. https://doi.org/10.1007/s00500-019-04580-5
21. Faheem M, Gungor VC (2017a) Energy efficient and QoS-aware routing protocol for wireless sensor network-based smart grid applications in the context of industry 4.0. Appl Soft Comput 68:910–922
22. Faheem M, Gungor VC (2017b) Capacity and spectrum-aware communication framework for wireless sensor network-based smart grid applications. Comput Stand Interfaces 53:48–58
23. Cui S, Zeng P, Song C, Wang Z, Li G (2022) Low-voltage distribution network topology identification based on constrained least square and graph theory. Soft Computing. https://doi.org/10.1007/s00500-022-07151-3
24. Guo S, Yan Y, Jiang J, Ma M, Bao W (2018) Network topology identification algorithm based on adjacency matrix. Power Syst Protect Control 46(12):50–56
25. Ahmadi A, Aghajari E, Zangeneh M (2022) High-impedance fault detection in power distribution grid systems based on support vector machine approach. Elect Eng. https://doi.org/10.1007/s00202-022-01544-1
26. Qu N, Zuo J, Chen J, Li Z (2019) Series arc fault detection of indoor power distribution system based on LVQ-NN and PSO-SVM. IEEE Access 7:184020–184028
27. Gunn SR (1998) Support vector machines for classification and regression. ISIS Tech Rep 14(1):5–16
28. Cristianini N, Shawe-Taylor J (2000) An introduction to support vector machines and other kernel-based learning methods. Cambridge University Press, Cambridge
29. Qian Z, Yaoquan Y (2012) Research on the kernel function of support vector machine. Electr Power Sci Eng 28(5):42–45
30. Babu NR, Mohan BJ (2017) Fault classification in power systems using EMD and SVM. Ain Shams Eng J 8(2):103–111

Chapter 5
Efficient Testing and Verification Methods

Introduction by the Editor

When the device is manufactured, lot of analysis is done to check if the modelled input output function resembles to the design on the silicon. Also, manufacturing defects needs to be checked to ensure design Is implemented as required. That's why this design verification step is a critical one in VLSI design. Chapter discusses about. Techniques used for efficient testing and verification method of the design which was fabricated. This chapter surveys various studies That can detect faults and minimise deviation from the Frontend model and the fabricated design. Such method discussed here is analysis and detection of open gate defects in redundant structures of FinFET SRAM cell, test set construct process for multi missing gate defects in reversible circuits, improved pseudo random fault coverage through inversions. In line is using automation and artificial intelligence to improve the process of verification for multiple FGPA platforms. It's a process which returns an error code by using neural networks and logic built-in self-test (LBIST). Point is to increase the effectiveness of test content to save time using techniques such as test content compression for scan architecture using modified shuffled shepherd optimization and speed up testing by using probabilistic merit indicators.

Multiple FPGA platforms are discussed at length in this chapter. FPGA usage can be summed up in techniques like simulations, emulations and prototyping. Aim of this is to validate maximum functionality of the chip in front end cycle. There are lots of areas in FGPA platforms which can be improved significantly. Electrical misreads are discussed in this chapter. This chapter as discusses on formulating satisfiability index and error localization effectiveness. The experiments mentioned in this chapter are based on signal clustering methodology, satisfiability-based post-silicon error localization and its scalability. Elaboration directs scope of improvement in selecting better trace signals for bit-flip localization for circuits under consideration. An in-depth discussion is also presented in this chapter related to

A. S. Shastri (ed.), *Optimization Methods in VLSI Design*,
https://doi.org/10.1007/978-981-95-2457-0_5

logic built in self-test. The study introduces deep learning in fault detection in combination and sequential circuits. Stuck-at and at-speed faults are also discussed in detail. Fault masking is to form a pattern of inputs which will translate to such values in the circuit that will mask the faulty lines and won't let the circuit functionality be affected.

Test point insertion (TPI) is one of the famous techniques to improve controllability and observability of the circuit which increases stuck-at and at-speed coverages. The study suggests use of only two hyperparameters to explore a big multi-dimensional parameter search space. These hyperparameters are training dataset size and number of neural networks used. The study reiterates the point that a hug training dataset can increase the training time and can at times lead to overfitting of the model which reduces the quality of output. This chapter also discusses the difference between Neural networks and problem-specific heuristics. It concludes that features like circuit structure, circuit paths and individual statement parameters can greatly affect network quality and causes of faults. Multiple missing gate faults in reversible circuits are discussed at length. The approach discussed helped create a test set for multiple consecutive missing gate faults in k-CNOT based reversible circuits. Probabilistic merit indicators are discussed in this chapter which help increases efficiency of simulation-based and meta heuristic test generation methods.

Machine Generated Summaries

Disclaimer: The summaries in this chapter were generated from Springer Nature publications using extractive AI auto-summarization: An extraction-based summarizer aims to identify the most important sentences of a text using an algorithm and uses those original sentences to create the auto-summary (unlike generative AI). As the constituted sentences are machine selected, they may not fully reflect the body of the work, so we strongly advise that the original content is read and cited. The auto generated summaries were curated by the editor to meet Springer Nature publication standards. To cite this content, please refer to the original papers.

Machine generated keywords: fault, test, test vector, vector, fault coverage, coverage, test set, verification, defect, generation, volume, scan, fault detection, soc., stuckat

Pre-Silicon Verification Using Multi-FPGA Platforms: A Review [1]

This is a machine-generated summary of:

Farooq, Umer; Mehrez, Habib: Pre-Silicon Verification Using Multi-FPGA Platforms: A Review [1]
Published in: Journal of Electronic Testing (2021)
Link to original: https://doi.org/10.1007/s10836-021-05929-1

If you want to cite the papers, please refer to the original.
For technical reasons we could not place the page where the original quote is coming from.

Abstract-Summary

"This has made pre-silicon verification even more pertinent as it can detect design faults prior to its roll out and can save companies a huge fortune."

"Pre-silicon verification now accounts for almost 70% of the total design effort and cost of modern digital systems."

"For pre-silicon verification, four techniques are commonly used namely simulation, emulation, virtual prototyping and FPGA-based prototyping."

"FPGA-based prototyping is unique in the sense it gives better speed and real world testing experience as compared to other pre-silicon verification techniques."

"We give a detailed survey of multi-FPGA prototyping."

"A survey of three different multi-FPGA platforms namely off-the-shelf, custom, and cabling platform is presented in this work."

"The survey is concluded with a discussion on the challenges faced by multi-FPGA prototyping and the research opportunities where work can be done for further improvement."

Introduction

"Design verification challenges in modern digital systems are huge and these challenges are increasing as the transistor size is shrinking."

"This is further exacerbated by the demands like parallel processing, multi threading; hence increasing the design complexity and making the design verification a hugely complex task."

"The discussion presented above highlights the importance of pre-silicon verification in the design process of digital systems."

"Authors in [2] present a comprehensive survey of formal methods used for hardware design verification."

"It can be seen here that the most of the aforecited review articles present different formal methods for verification of digital systems except for [3] where some simulation and emulation methods are discussed for memory verification."

"We present a detailed discussion on four different pre-silicon hardware verification techniques."

"The main contributions of this work are summarized as follows: A detailed review of four pre-silicon verification techniques is presented."

Pre-silicon Verification Techniques

"Simulation-based verification technique offers better scalability as compared to other verification techniques."

"ModelSim [4] is another simulation-based verification solution that is provided by Mentor Graphics."

"Just like simulation, emulation based verification solutions are also offered by various vendors."

"Real time verification of the design of an SoC using simulation or emulation is hard because of the complexity of SoC. FPGA-based prototyping, allows to run the RTL of an SoC in real time while giving cycle-accurate and bit-accurate information about the design under test."

"The verification of the design can be performed with real, external world interfaces like DDR, PCI, USB etc Furthermore, the execution speed offered by FPGA-based prototyping is far superior than simulation and emulation based verification."

"Although FPGA-based prototyping offers advantages that are unique to it, there are some disadvantages also which are associated with this verification technique."

Multi-FPGA Platforms

"The number of FPGAs in a multi-FPGA platform vary as they depend upon the complexity of the design under consideration."

"A hardwired off-the-shelf multi-FPGA platform is a pre-designed and pre-fabricated board."

"In the custom multi-FPGA board, the platform is built according the requirements of design under consideration."

"Cabling multi-FPGA platform is comparatively new as compared to off-the-shelf and full custom multi-FPGA boards."

"In a cabling multi-FPGA platform the FPGA boards are already available and they are connected to each other through cables and connectors."

"Just like off-the-shelf platform, cabling platform is readily available as the FPGA boards are already fabricated and like custom FPGA boards, it also offers the option of customizing the connections between FPGAs using connectors and cables."

"For cabling platform, the availability time is smaller as compared to custom multi-FPGA board."

"Cabling multi-FPGA platform offers highest flexibility as the cable connections can be modified for different designs."

Back End Flow for Multi-FPGA Platforms

"It is evident from the discussion presented in this section that partitioning is an important and challenging step of the multi-FPGA prototyping flow."

"The number of signals (also termed as cut-nets) traversing different partitions are more than the available I/Os between different FPGAs of a multi-FPGA board."

"When designs are partitioned for multi-FPGA prototyping, there can be two different types of cut-nets."

"Once the multi-FPGA partitioning and inter-FPGA routing steps are complete, the partitions of the designs are given to vendor specific tools for intra-FPGA synthesis, placement, and routing."

"Authors in [5], use the same approach to perform comparison between off-the-shelf and custom-multi-FPGA platforms."

"Authors in [6] present an algorithm to automate the process of designing a cabling platform for multi-FPGA prototyping and they have claimed to reduce the design time significantly."

Challenges in Multi-FPGA Prototyping and Opportunities

"In this section, we present detailed discussion on the challenges faced by multi-FPGA prototyping both from platform and back end flow perspectives."

"Inter-FPGA delays are becoming larger and it has led to the decreasing performance of the designs that are prototyped using FPGAs of newer generations."

"The current flows of custom and cabling multi-FPGA platforms are limited in the sense that they consider all the FPGAs in custom and cabling multi-FPGA platforms are of same type and same family."

"Existing flows for custom and cabling platforms do not cater the FPGAs from different vendors or even different FPGA families of the same vendor."

"Considering only a single type of FPGA in the multi-FPGA platform might lead to under utilization of some resources and may lead to lower than the best possible performance of the design."

Conclusion

"We have reviewed and discussed three different multi-FPGA platforms namely off-the-shelf, custom, and cabling multi-FPGA platform."

"Their performance is poor when compared to custom and cabling multi-FPGA platforms."

"Custom multi-FPGA platforms have very good performance."

"As far as cabling platform is concerned, it is comparatively new and if we look at it in terms of availability and performance, it lies in between off-the-shelf and custom multi-FPGA platform."

"Apart from various advantages of multi-FPGA platforms, they are also faced by a number of challenges like non-standardized design flow, increasing logic to I/O ratio, lack of standard benchmarks and comparison metrics, etc A discussion on the challenges faced by multi-FPGA prototyping reveals that still a lot needs to be done in this domain."

SAT-Based Silicon Debug of Electrical Errors Under Restricted Observability Enhancement [7]

This is a machine-generated summary of:

Kumar, Binod; Fujita, Masahiro; Singh, Virendra: SAT-based Silicon Debug of Electrical Errors under Restricted Observability Enhancement [7]

Published in: Journal of Electronic Testing (2019)

Link to original: https://doi.org/10.1007/s10836-019-05830-y

If you want to cite the papers, please refer to the original.

For technical reasons we could not place the page where the original quote is coming from.

Abstract-Summary

"Silicon debugging of integrated circuits is exacerbated by the lack of golden responses, highly restricted observability and irreproducible nature of bugs."

"Debug engineers need to develop better methods that can assist in error localization at lower level(netlist) granularity."

"This paper revisits methodologies to debug electrical errors through satisfiability(SAT) solving under a limited visibility environment."

"We propose various SAT formulations and analyze their efficacy in error localization for a variety of benchmark circuits."

"We aim at minimizing the overhead associated with signal tracing while maintaining the error localization efficacy."

"We address scalability concerns in SAT solving through partitioning of large error traces."

Introduction

"Restricted observability and controllability of the internal design states along with the lack of failure repeatability and scarcely available golden responses are some of the major obstacles during the post-silicon validation and debugging [8, 9]."

"This paper proposes a graph-based signal tracing methodology to address the signal selection problem and then utilize satisfiability-based debugging of the obtained post-silicon signal dumps."

"Since post-silicon execution has extremely limited visibility, obtaining constraints for solving the SAT instances is a challenging task and involves trade-offs of debug time, deployed design-for-debug features and required off-chip dumping."

"We also discuss the scalability issues of SAT-based debugging under post-silicon environment for electrical error scenarios."

"The major contributions of the paper are: a methodology is proposed for tracing of signals for collecting debug data through different observability enhancement schemes. A satisfiability-based debugging methodology is presented for handling large error traces. Detailed evaluation metrics are proposed to measure the efficacy of SAT-based electrical error debug under limited post-silicon observability."

Background

"Trace buffers are widely utilized as a design-for-debug feature for storing a selected group of signals for a specified number of clock cycles, which can then be dumped off-line for debugging."

"The metric of state restoration fails to capture other important issues such as ease in error localization/debugging with the assistance of traced and restored signal states."

"The contents of these trace signals(flip-flops) are continuously dumped into the trace buffers for six clock cycles."

"In a static signal tracing scheme, only two signals can be dumped for 6 clock cycles, whereas in this scheme, four signals (flip-flops) meaning that all the signals are dumped in alternate cycles."

"Since larger number of internal signals can be covered for signal tracing with this observability enhancement scheme, error localization can be more effective."

"For error localization, we need to compute the UNSAT core(s) after forcing constraints (obtained from the error trace) on the CNF expression of unrolled netlist."

Proposed Signal Clustering Methodology

"The proposed grouping/clustering of the signals is aimed at minimizing the number of trace signals required for SAT-based error localization."

"The proposed methodology consists of two parts—first, the netlist connections are represented by a graph (G) where the flip-flops are represented by nodes, second, a merging procedure is followed to cluster the candidate signals and finally select trace signals."

"Under such scenario(s), an unequal number of flip-flops(trace signals) can be chosen from each of the three obtained clusters or from two of them."

"Once the clusters are formed, the final decision regarding the selection of trace signals is done based on graph centrality measures."

"An interesting proposition is to select final trace signals by simply utilizing in-degree centrality measure instead of cluster formation and then subsequent selection."

SAT-Based Post-silicon Error Localization

"The complete SAT-based localization analysis is then performed for these smaller error traces ($eTrace_j$) leading to the unrolling of the circuit for only 100 cycles compared to the case when the netlist needs to be unrolled for a total of 1000 cycles (which is complete error trace length)."

"To measure the quantity of exact error localization($C \in Su_{cc}$, $F \in Su_{fc}$) for bit-flip errors, we define the following metric: The above metric gives the number of iterations (out of a total of N_E experiments) in which both the suspect flip-cycle and flip-flop are obtained."

"Similar to the bit-flip error case, the following metric is defined to measure the efficacy of SAT-based stuck-at error localization (where $net \in Su_{sa}$ and net_i is the wire/net in the netlist which is stuck-at): Another metric worth consideration is the total number of suspect nets obtained after analysis of UNSAT cores in the error injection experiments: It is obvious that lower the values of ϕ_9, better is the stuck-at error localization."

Experimental Results and Observations

"For instance—we observed a maximum decrease of 90% in ϕ_9 values for b21 circuit when the tracing configuration is changed from scan2 to scan4 for bit-flip errors."

"Major observations regarding the utility of different SAT formulations can be summarized as follows: Trace signal selection has no impact on localization with B_2."

"B_4 provides lower average flip-cycle (ϕ_3) compared to B_3 which means that the flip-cycle candidates are closer to the actual flip-cycle with former formulation compared to the latter formulation (B_3). For all the circuits, the number of UNSAT attempts with B_2 is less that of other formulations which means that error does not propagate to primary outputs (PO) in those cases."

"B_1, B_3 and B_4 always result in UNSAT.4.In cases of inexact localization, compared to other formulation, B_4 provides lower topological distance (ϕ_7) of flip-flop suspects from the actual flip-flop."

Addressing Different Scalability Issues in SAT-Based Error Localization

"In spite of scalability limitations of SAT solvers, we believe that SAT-based debugging methodology in post-silicon environment can be applied to any large circuit provided the length of error trace is reasonably smaller."

"Time spent in the design unrolling stage is not very critical for applying the methodology to large circuits or applying it to long error traces."

"We observed a linear increase in the time required for unrolling of the design netlist when the error trace length (the number of clock cycles) is doubled."

"Time and memory usage increase is linear with respect to error trace length also."

"One of the restrictions of this debugging methodology is the requirement of unrolling the netlist into time-frames equal to the length of an error trace."

"We did not observe any appreciable difference in the time and memory usage across all the benchmark circuits for SAT solving when the trace signal selection technique is varied."

Discussions on Previous Silicon Debug Methods and Future Enhancements

"This methodology can be utilized in conjunction with the proposed technique for effective post-silicon error localization as the trace signal selection can be enhanced through probability analysis."

"We identify following avenues to improve the error localization for some of the circuits: (1) assigning proper primary inputs from the design knowledge (i.e., hierarchical analysis of RTL). (2) obtaining a useful initial state of the design netlist (i.e., signal states of all flip-flops corresponding to a cycle before the beginning of error trace) from design simulation to assist in SAT solving. (3) setting intermediate controls signals to their appropriate values to obtain proper UNSAT cores(s)We believe that above enhancements would play a more important role in the error localization for circuits of opencore and ITC'99 suites than a bit-flip detection driven trace signal selection (like the one proposed in [9])."

Conclusion

"This paper proposed a signal tracing methodology for satisfiability based error localization in the post-silicon environment."

"A grouping based methodology is presented for signal selection and it is used with different observability enhancement schemes to obtain a smaller execution trace."

"The error trace obtained from chip execution is forced on the time-frame unrolled circuit for satisfiability solving as constraints."

"Experiments on a variety of benchmark circuits show that SAT-based debugging methodology succeeds in the post-silicon environment for two kinds of error scenarios."

"The graph-based signal grouping methodology can be enhanced to select better trace signals for bit-flip localization for those circuits."

Automated Design Error Debugging of Digital VLSI Circuits [10]

This is a machine-generated summary of:

Moness, Mohammed; Gaber, Lamya; Hussein, Aziza I.; Ali, Hanafy M.: Automated Design Error Debugging of Digital VLSI Circuits [10]

Published in: Journal of Electronic Testing (2022)

Link to original: https://doi.org/10.1007/s10836-022-06020-z

If you want to cite the papers, please refer to the original.

For technical reasons we could not place the page where the original quote is coming from.

Abstract-Summary

"We propose a new approach utilizing deep learning for fault detection (FD) of combinational and sequential circuits in a type of stuck-at-faults."

"The goal of the proposed semi-supervised FD model is to avoid the search space explosion problem by taking advantage of unsupervised and supervised learning processes."

"The supervised process tends to describe rules of classification that are applied to the reduced features for detecting different stuck-at faults within circuits."

"The FD model proposes good performance in terms of running time about 187×compared to other FD algorithm based on SAT solvers."

"It is compared to common classical machine learning models such as Decision Tree (DT), Random Forest (RF) and Gradient Boosting (GB) classifiers, in terms of validation accuracy."

"The results show a maximum validation accuracy of the feature extraction process at 99.93%, using Deep sparse autoencoder for combinational circuits."

"The fault detection process delivers around 99.6% maximum validation accuracy for combinational circuits from ISCAS'85 and 99.8% for sequential circuits from ISCAS'89 benchmarks."

"The proposed FD model has achieved a running time of about 1.7×, compared to DT classifier and around 1.6×, compared to RF classifier and GB machine learning classifiers, in terms of validation accuracy in detecting faults occurred in eight different digital circuits."

"The proposed model outperforms other FD models, based on Radial Basis Function Network (RBFN), achieving 97.8% maximum validation accuracy."

Introduction

"For detecting logic faults in the case of erroneous digital circuits, most of these approaches are based on detecting different subsets that give valuable meaning to designers, such as minimal correction subsets (MCSes) and minimal unsatisfiable subsets (MUSes)."

"Despite most approaches based on ML are focused on locating design faults at a post-silicon stage, detecting faults in a pre-silicon stage is more important to make sure of all components in digital circuits before going deeper into the design flow [11]."

"The conducted research introduces the following contributions that are more promising than SAT: The search space explosion problem in FD approaches based on formal methods can be solved using counterexample resulting from failed verification passed to semi-supervised learning model for classifying stuck-at-faults, which might be causes for failing verification."

"Exact debuggers using Max-SAT approach for detecting the main reason of faults existed in circuit is replaced by ATALANTA tool for generating multiple patterns for each stuck-at-fault with high fault coverage."

Background

"The proposed algorithm is focused on detecting stuck-at-faults that can be mapped to gate replacement faults in logical circuits as proposed in [12], proving set of corollaries for describing mapping from stuck-at-faults into design fault model domain."

"The method of mapping Stuck-at faults into gate-replacement faults is concluded from the following proof in [12]: For detecting design errors in the implementation at any gate in a given circuit, it is sufficient to apply a pair of test patterns which detect the stuck-at one and stuck-at zero at one of the gate inputs."

"The following set of corollaries describes the mapping process from stuck-at fault model to the design error domain."

"Data-driven fault diagnosis that is implemented based on machine learning models (such as support vector machine (SVM), neural network (NN) and fuzzy logic) is more common because of the data analytical methods in them."

State of the Art for Fault Debugging

"The following two subsections discuss different approaches for fault detection based on machine learning or deep learning and fault detection and localization based on SAT solvers."

"This method exploits the power of autoencoder in strong low-dimensional non-linear representations for detecting rare events and the strength of LSTM in time series learning ability for the fault diagnosis."

"Once the autoencoder detects a fault, the dependencies between various time steps of sequential data can be learned by the LSTM network to identify the types of faults."

"The main problem of SAT-based Fault detection is multiple SAT solver calls."

"Of solving the CNF equation (which is in large size after injecting debug circuits) by one of SAT solvers, our model can learn the connections between lines from the data set which contains test patterns with corresponding correct outputs and the corresponding fault mask."

The Proposed Implementation

"This can be achieved by defining features as all inputs and output pins of digital circuits with the complete test patterns which can detect all stuck-at-faults as features of our model."

"As we attempt to detect faults on a given circuit without implicitly injecting the structure of a circuit due to its large size, we use test patterns with its correct responses and their fault mask for forming a model that can approximate the connections between lines which are faulty lines due to single stuck-at fault."

"The validation accuracy of the autoencoder means that the autoencoder can extract the most important part of test patterns and faulty-free responses of a circuit for detecting all possible stuck-at fault."

"The meaningful inputs and outputs can be extracted from all test patterns for helping model to detect faults quickly and solve the search space explosion problem occurred in traditional fault detection methods especially faults occurred in large-sized digital circuits."

Experimental Results

"After combining Deep AE with Classifier layer using sigmoid activation function for multilabel classification as detecting multiple faults in digital circuits, our model is trained on GPU for classifying multiple stuck-at faults in digital circuits, using the latent space extracted from previous step."

"With the growing size of inputs and outputs of sequential circuits which causes search space explosion in other approaches, the proposed model accomplishes around 99.7% as validation accuracy to detect possible faults in "s13207.1", which contains 1,490 input and output port."

"The performance of proposed FD-based on DL outperforms SAT fault detection, as a result of dispense of searching process and multiple calls of SAT solvers and construct DL model of IC design interpreting possible multiple stuck-at-faults by extracting features from test vectors and classifying multiple faults."

Conclusion

"The main contribution of this work is to solving the search space explosion due to large size circuits by designing a semi-supervised model that is mainly based on Deep Sparse Auto-encoder for dimensionality reduction of features represented correct IC design."

"Different types of autoencoders (normal AE, Deep Sparse AE and stacked sparse AE) as unsupervised model, were implemented in order to find the latent representation of test patterns with their correct response for the next phase of fault classifier."

"Our model can extract the main connections in a given circuit which leads to multiple equivalent faults."

"The main advantage of our model is avoiding multiple calls of SAT solver and the reduction of search space problem for generating a specific MUS that can extract faults."

Applying Artificial Neural Networks to Logic Built-in Self-Test: Improving Test Point Insertion [13]

This is a machine-generated summary of:

Sun, Yang; Millican, Spencer K.: Applying Artificial Neural Networks to Logic Built-in Self-test: Improving Test Point Insertion [13]

Published in: Journal of Electronic Testing (2022)

Link to original: https://doi.org/10.1007/s10836-022-06016-9

If you want to cite the papers, please refer to the original.

For technical reasons we could not place the page where the original quote is coming from.

Abstract-Summary

"This study applies artificial neural networks (ANNs) to increase stuck-at and delay fault coverage of logic built-in self-test (LBIST) through test point insertion (TPI)."

"This study demonstrates how to train an ANN to evaluate test points (TPs) and demonstrates a substantial decrease in TPI computation time compared to existing heuristics while delivering comparable stuck-at and delay fault coverage."

Introduction

"High-quality tests may need to be applied in-the-field to ensure continuing circuit reliability: aggressive technology scaling accelerates post-manufacturing degradation and the occurrence of intermittent soft errors [14], and like pre-delivery tests, tests must detect (and possibly remedy) these in-field defects."

"The logic circuit industry uses LBIST to apply manufacturing tests since LBIST can obtain significant fault coverage while decreasing test application time, which in turn reduces manufacturing test costs."

"Keeping LBIST quality high is challenging under aggressive technology scaling since logic circuit complexity is growing faster than design-for test (DFT) techniques can remedy."

"One particular technique, test point (TP) insertion (TPI) [15–19], is the focus of this article, since TPI can be used in conjunction with other techniques to further increase LBIST quality."

"New computing methods, such as artificial neural networks (ANNs), can increase algorithm efficiency and keep LBIST quality high."

"This article demonstrates how ANNs can increase TP quality when applied to TPI while drastically decreasing algorithm runtime."

Background & Motivation

"A typical example of an RPR fault is the output of a 64-bit wide OR gate (e.g., a global error-detecting circuit which polls 64 individual error lines) stuck-at logic-1."

"To make RPR faults easier to detect, TPs make circuit lines either (1) easier to control or (2) easier to observe."

"Although algorithms typically model TE and observe points as pins for simplicity (similar to how scannable latches are typically modeled as circuit inputs and outputs by test generators), implementing them as circuit pins is impractical given their high cost."

"Many studies demonstrated TPs effectively testing stuck-at faults, but control points can lower delay fault coverage when TPI targets stuck-at faults."

"Control points may be the best TP choice for some circuits, since some faults (stuck-at or delay) may require forced circuit lines, which observe points cannot create."

ANN for Stuck-at Fault TPI

"The ANNs evaluate TPs using circuit probability information, i.e., the same controllability-observability program (COP) values [20] that many other TPI methods use [16, 17, 19], but unlike other TPI methods, the ANNs do not require recalculating this information when evaluating a TP."

"This study attempted to collect training data through the direct application of vectors to circuits, but in reasonable overnight runs, only hundreds of training samples were collected, and the resulting ANNs selected TPs that consistently failed to increase fault coverage regardless of training effort."

"The first training speedup technique is to apply fault simulation to subcircuits with assistance from circuit probability calculations (i.e., COP values)."

"Although this technique significantly decreases training data generation time, the technique may hinder the ANNs' ability to select high-quality TPs since subcircuit controllability and observability values are not one hundred percent accurate [20]."

ANN for Delay Fault TPI

"The training data collection program uses the transition delay fault (TDF) model in lieu of the stuck-at fault model."

"Like the stuck-at fault-qualifying ANNs, the delay fault-qualifying ANNs train under various ANN hyperparameters to minimize error."

"The hyperparameters of these ANNs may not be the same as the stuck-at fault-qualifying ANNs, but to allow for a fair comparison of the two, the following sections' evaluations use the same amount of training time for both ANNs."

ANN Training Experiments

"Given the vast parameter exploration space, this article limits itself to showing the exploration of two hyperparameters: training data set size and the number of ANN neurons."

"Providing too much training data increases training time and can degrade ANN quality through "overfitting" [21]."

"Training requires more time when using more data; this is because minor changes in ANN weights and biases are more likely to create error among more training data samples."

"Using more training data can increase the error, since finding ANN weights that satisfy all training samples becomes more difficult."

"Since using 20,910 samples doubled training time with a marginal impact on error, future experiments will use ANNs trained with 10,457 training data samples."

"10,457 data samples train the stuck-at fault targeting, control-1 TP-evaluating ANN with all other parameters matching the previous experiment."

ANN Vs. Heuristic TPI Experiments

"The goal of this study is to observe the impact of replacing critical subroutines with creatively-trained ANNs, thus in lieu of industrial-designed TPI tools with many difficult-to-isolate sources of CPU time, the conventional stuck-at fault TPI method used for comparison is from Tsai and others [19]."

"This study shows the efficacy of using ANNs in lieu of circuit-wide heuristics can have on a TPI time and quality, thus it is logical to presume applying this study's methods to the heuristics of more contemporary TPI—like Maghaddam and others's [22]—will yield similar trends in CPU time reduction and fault coverage improvement."

"This limits larger circuits to a single TP, but industrial tools and contemporary TPI algorithms will presumably see similar TPI time and fault coverage trends when inserting more TPs."

Conclusions & Future Directions

"This article extended work done in previous studies [18, 23] by analyzing features relevant to LBIST tests (i.e., the number of vectors in the test), explored a more relevant label for evaluating a TP in the context of an LBIST test, and made a thorough exploration of the impact ANN complexity has on TPI outcomes."

"The authors anticipate continuing this work by finding features and training methods that further increase TP quality while inserting TPs under additional constraints, especially methods that consider test power while using nuanced delay fault models."

"This study chose features used by existing heuristics, thus making a fair comparison possible; however, additional features (e.g., circuit structure, critical paths, etc) may drastically increase the ANN's quality and reveal nonintuitive causes of fault coverage degradation."

Analysis and Detection of Open-Gate Defects in Redundant Structures of a FinFET SRAM Cell [24]

This is a machine-generated summary of:

Champac, Victor; Mesalles, Javier; Villacorta, Hector; Vargas, Fabian: Analysis and Detection of Open-gate Defects in Redundant Structures of a FinFET SRAM Cell [24]

Published in: Journal of Electronic Testing (2021)

Link to original: https://doi.org/10.1007/s10836-021-05955-z

If you want to cite the papers, please refer to the original.

For technical reasons we could not place the page where the original quote is coming from.

Abstract-Summary

"FinFET technology is used in leading high-performance/power-efficient electronic products."

"It is required to study the behavior of defects in FinFET-based SRAM memories and develop new test strategies for those not covered by conventional test strategies based on CMOS fault modeling."

"This paper study open-gate defects affecting only one of the parallel fins in the driver transistors of the memory cell."

"The behavior of these defects is studied for the hold, read and write operations using realistic defect models."

"By using a short write time test, the detection of these defects is investigated."

"The detection probability of these defects can be further enhanced using a higher power supply voltage."

Introduction

"New defects may occur in FinFET circuits as new technologies are incorporated in their manufacturing process and the use of multi-fin/multi-finger structures."

"Conventional CMOS fault models can cover some defects, but some defects are specific to FinFET technology."

"Bhoj and others [25] developed fault models for opens and shorts in FinFET gates."

"Liu and others [26] studied different types of defect in FinFETs such as fin stuck-on, stuck open, and gate oxide short."

"Harutyunyan and others [27, 28] have investigated the detection of manufacturing defects such as opens and shorts in FinFET memory cells."

"The impact of parameter variations on the fault coverage of FinFET memory cells when dealing with FinFET-specific faults are studied in [29]."

"Chen and others [30] have studied FinFET specific faults in a FinFET-based SRAM circuit."

"The behavior of open-gate defects in redundant structures in FinFET memory cells is studied."

Open-Gate Defects and Process Variations

"As the value of this charge can not be predicted, the behavior of the memory cell is analyzed for an assumed range of possible initial voltages at the floating node."

"The trends of the results remain for other values of the trapped gate charge."

"Middle-of-line (MOL) interconnection is used for local routing within a cell in advanced technology nodes [31]."

"CA layer is used to connect the drain/source of the fins, and the CB layer is used to connect the gate fingers."

Performance Analysis in a FinFET SRAM Cell with Open-Gate Defects

"The behavior of the defective memory cell is analyzed for the following operations: (a) hold, (b) read, and (c) write."

"The defect-free cross-coupled inverter of the memory cell is not shown."

"For hold-1 operation, the memory cell presents deviation from the defect-free behavior."

"Because both parallel fins of the driver transistor of INV2 are defect-free, the memory cell is correctly read as the transient simulation shows."

"For read-1 operation, the memory cell presents deviations from the defect-free behavior."

"Because the discharge of Q is done through the access transistor and writing circuitry, the memory cell can perform the write operation."

"The previous analysis has shown that the defective memory cell can make its operation functionalities correctly."

"The functional redundancy in the drivers of the memory cell allows correct hold and read/write operations."

"The defective memory cell presents a degraded SNM, which further reduces due to process variations."

Detection of Open-Gate Defects Located in Redundant Structures of an SRAM Cell

"The memory cell behaves as follows: Bad write of the defect-free memory cell."

"Unlike conventional test, this technique is based on the fact that a defect-free memory cell fails the write operation, but a defective memory cell can perform the write operation correctly."

"The write time assures a good write operation for the defective memory cell and a bad write operation for the defect-free memory cell."

"The transfer curves of the defect-free and defective memory cells may suffer shifts due to the parameter process variations."

Application of the Short Write Time Test

"The minimum write time of a defect-free memory cell may take different values due to process variation of the parameters."

"The short write time test technique is based on a bad write of a defect-free memory cell and a good write of a defective memory cell."

"A short write time DFT circuitry to test open defects in memristor-based memories has been proposed in [32]."

"The second objective of our work is to analyze the test efficiency of the short write time test to detect open-gate defects in a FinFET SRAM cell."

"Even more, conditions at nominal and under process variations are given to choose a proper write time to assure a bad write of the defect-free memory cell and at the same time to assure a good write of the defective memory cell."

Conclusion

"The behavior of open-gate defects located in redundant structures of a FinFET SRAM cell has been investigated."

"The open-gate defect affects only one of the parallel fins of the driver transistors of the memory cell."

"A higher power supply voltage increases the detection probability for these open-gate defects."

An Efficient Test Set Construction Scheme for Multiple Missing-Gate Faults in Reversible Circuits [33]

This is a machine-generated summary of:

Handique, Mousum; Deka, Jantindra Kumar; Biswas, Santosh: An Efficient Test Set Construction Scheme for Multiple Missing-Gate Faults in Reversible Circuits [33]

Published in: Journal of Electronic Testing (2020)

Link to original: https://doi.org/10.1007/s10836-020-05855-8

If you want to cite the papers, please refer to the original.

For technical reasons we could not place the page where the original quote is coming from.

Abstract-Summary

"The proposed work provides a scheme for generating the complete test set for detecting the single and any number of consecutive multiple missing gate faults in k-CNOT based reversible circuits."

"The complete test set generation method is twofold."

"A local test pattern is applied to each level of k-CNOT gates and the reverse simulation method is used for identifying all the possible Single Missing Gate Faults (SMGFs)."

"Using the complete test set for SMGFs and based on the structure of the k-CNOT based circuit, a test set is formulated."

"The generated test set is capable of detecting all the MMGFs and as well as the SMGFs in reversible circuits."

Introduction

"The properties of controllability and observability of reversible circuit help to generate the test set to detect faults and we propose methods to generate the minimal test set to detect single missing gate faults and consecutive multiple missing gate faults in such circuits."

"An ATPG (Automatic Test Pattern Generation) method is proposed to generate the complete test set for detecting the single and consecutive multiple missing gate faults (MMGF) in the reversible circuits."

"The main contributions of this work are as follows: (1) ATPG algorithms are proposed to generate the complete test set for SMGFs and MMGFs, and extract the minimal complete test set for detecting these faults in the reversible circuits. (2) Correlate the fault coverage range with other fault models such as stuck-at faults, partial missing gate faults, and appearance crosspoint faults."

Background

"And 2011, Rahaman and others [34, 35] proposed a DFT method for an (n × n) reversible circuit by adding only one extra line along with duplication of each k-CNOT gate to get a universal test set of size (n + 1), which is sufficient to detect all SMGFs along with other faults like RGFs and PMGFs."

"Zhang and others [36] proposed an ATPG method using the concept of Boolean satisfiability (SAT) for generating the complete test set, which can detect the single missing control fault (SMCF), single additional control fault (SACF), and single missing-gate fault (SMGF) in reversible circuits."

"We propose an ATPG method to determine the complete test set for detecting all SMGFs and MMGFs using the reversible circuit properties (controllability and observability) without changing the structure of the circuit and the time complexity of the method is $O(N^2)$ with N number of gates in a k-CNOT circuit."

Proposed Method

"The test set, TS_{SMGF}, generated using the proposed method detects all single missing gate faults in a given reversible circuit with N-gates."

"The test set TS generated using the proposed method detects all single and multiple missing gate faults in a given n-input k-CNOT based reversible circuit."

"The test set TS_{MIN} is a minimal complete test set for detecting all the single and multiple missing gate faults in a given n-input k-CNOT based reversible circuit."

"The complete test set TS for SMGFs and MMGFs satisfies the requirement for detecting most of the SAFs in a given reversible circuit. Missing gate fault and PMGF model: The detection criteria for a PMGFs is that the missing control input is set to logic value 0 and we assign the logic value 1 to all other control inputs."

Experimental Results and Discussions

"The authors in [37] proposed a greedy and BDD based covering method for generating the minimal test set for detecting all possible cases of two consecutive missing gate faults."

"The reduction of test set by our proposed test pattern generation method to detect MMGFs is more than or equal to 50% for more than 77% of the benchmark circuits that are used in the experiment."

"The authors in [38] proposed two ATPG approaches to generate the complete test set using the naive-based and SAT-based approach for detecting the SMGF along with SMCF (Single Missing Control Faults), and PMGF in reversible circuits."

"Experiments are performed to check the fault coverage range of other fault models such as stuck-at faults (SAFs), partial missing gate faults (PMGFs) and crosspoint faults using the complete test set TS generated by our proposed method."

Conclusion

"This paper presents a scheme to generate the complete test set for detecting single and any number of consecutive multiple missing gate faults in the k-CNOT based reversible circuits."

"The complete test set generation method is twofold."

"The fault coverage ranges for the SAFs, PMGFs and appearance crosspoint faults with our generated complete test set are also analyzed and checked using experimental results."

"By looking into the correlation between different fault models, there is a possibility to reconstruct the generated complete test set to cover these fault models such that all the possible faults in reversible circuits can be detected by a single test set."

Speed-Up in Test Methods Using Probabilistic Merit Indicators [39]

This is a machine-generated summary of:

Fooladi, Mahtab; Kamran, Arezoo: Speed-Up in Test Methods Using Probabilistic Merit Indicators [39]

Published in: Journal of Electronic Testing (2020)

Link to original: https://doi.org/10.1007/s10836-020-05871-8

If you want to cite the papers, please refer to the original.

For technical reasons we could not place the page where the original quote is coming from.

Abstract-Summary

"Deterministic test generation methods are time consuming and, this has led to emergence of simulation-based approaches."

"The basis of simulation-based methods is to propose a number of test vectors, evaluate the efficiency of the proposed vectors, and accept or reject them."

"In these methods, the efficiency of each test vector is traditionally evaluated based on fault coverage of that vector."

"The proposed measure can be used in simulation-based and meta-heuristic test generation methods to achieve speedup."

"In order to indicate efficiency of the proposed probabilistic measure, two test generation methods that utilize this measure are suggested and exploited to generate test for ISCAS85 benchmark circuits."

"These test generation methods are called PMI_PRUNE and PMI_GA."

Introduction

"Deterministic methods provide the possibility of obtaining high fault coverages but these methods are time consuming and furthermore require a test compaction phase to reduce the number of test vectors generated."

"A set of test vectors are generated randomly or based on some sort of heuristic and then, by forward analysis of the circuit, the best vectors, in terms of their fault coverage, are selected and placed in the final test set."

"Test generation time in these methods is shorter than those based on genetic algorithm, but the test quality is lower."

"In simulation-based test generation methods, fault coverage is the main criterion used to assess efficiency and merit of test vectors."

"Our main contribution in this paper is proposing a new probabilistic fault injection method, proposing and evaluating three probabilistic measures to substitute fault coverage, and adapting pruning-based and GA-based test generation methods to effectively utilize the best proposed probabilistic indicator."

Probabilistic Circuit Analysis

"D means that the signal value must be 1 but it is changed to 0, because the line has an activated stuck-at-0 fault or effect of an activated fault elsewhere in the circuit, has propagated to this point and changed the correct value of the signal from 1 to 0."

"The most important advantage of probabilistic four-value system is that it allows simultaneous injection of many faults into a circuit."

"Instead of injecting faults one by one and performing one round of simulation per fault, the probabilistic system enables us to generate a signature from cumulative effect of all faults on the primary outputs, in one probabilistic simulation round."

"While, in probabilistic fault simulation, all faults are injected simultaneously, and one round of probabilistic calculation is performed."

Proposed Merit Indicators and Their Evaluation

"To evaluate the correlation of the proposed PMI merit indicators, we have generated ten sample sets of random test vectors (each sample including 2000 random test patterns) for each circuit."

"We have calculated correlation of PMI measures with fault coverage for each sample set at several different values of r in the range of 10^{-10} to 6×10^{-1}."

"To have a better understanding on the efficiency of PMI_SUM, we have used another method called fault sampling to achieve an estimate for fault coverage of a test vector."

"Comparison of PMI_SUM and fault sampling method shows that PMI_SUM has relatively higher correlation (in most circuits, even for sampling rate of 10% which has the highest correlation) with much greater speedup (even in sampling rate of 1% which has the highest speedup)."

[Section 4]

"In Section 3, we proposed a probabilistic measure called PMI_SUM which can estimate efficiency of test vectors much faster than the traditional fault coverage measure with an acceptable approximation."

"To show and evaluate effectiveness of PMI_SUM, we have proposed and adapted two different combinational test generation methods that utilize PMI_SUM instead of fault coverage metric."

"Each test vector is then applied to the inputs of the circuit and by performing probabilistic fault simulation, PMI_SUM is calculated for that test vector."

"If PMI_SUM of this test vector is higher than the threshold value (th_PMI_SUM), that test vector is accepted and added to the final test set."

"To accelerate GA-based ATPGs, we propose to use PMI_SUM as an approximation of fault coverage, to determine fitness of test vectors."

"PMI_SUM is then calculated for all test vectors in the new generation."

Experimental Results

"This section discusses two proposed test generation methods, PMI_PRUNE, and PMI_GA."

"To implement the proposed test generation methods, C ++ language and LPSim simulation engine have been used."

"In order to evaluate PMI_PRUNE and PMI_GA, a test set has been generated for each circuit of ISCAS85 benchmark, by these methods."

"The quality of these test sets (in terms of fault coverage and the number of test vectors), as well as test generation time, is evaluated and compared with results of baseline methods that traditionally use fault coverage (DMI_PRUNE, and DMI_GA)."

"Shows the number of fault detected, #TVs denotes the number of test vectors, and t_{norm} is the normalized test generation time."

"The results confirms that using the proposed probabilistic measure, PMI_SUM, instead of fault coverage, we can decrease test generation time by a slight increase in the number of test vectors."

Discussion and Conclusion

"Simulation-based methods use fault coverage as a precise measure to decide about efficiency of test vectors."

"Our investigations show that among the proposed measures, PMI_SUM has a higher association with fault coverage."

"Our evaluations also demonstrate that PMI_SUM, on average, can be calculated 782× faster than fault coverage."

"In order to confirm this claim, this paper proposes two test generation methods, PMI_PRUNE, and PMI_GA, which exploit the probabilistic measure PMI_SUM instead of fault coverage."

Improved Pseudo-Random Fault Coverage Through Inversions: A Study on Test Point Architectures [40]

This is a machine-generated summary of:

Roy, Soham; Stiene, Brandon; Millican, Spencer K.; Agrawal, Vishwani D.: Improved Pseudo-Random Fault Coverage Through Inversions: a Study on Test Point Architectures [40]

Published in: Journal of Electronic Testing (2020)

Link to original: https://doi.org/10.1007/s10836-020-05859-4

If you want to cite the papers, please refer to the original.

For technical reasons we could not place the page where the original quote is coming from.

Abstract-Summary

"This article analyzes and rationalizes the capabilities of inversion test points (TPs) when implemented in lieu of traditional test point architectures."

"Delay faults must be targeted by LBIST, but delay faults can be masked when using control-0/1 (i.e., traditional) TP architectures."

"This study extends work previously presented in the North Atlantic Test Workshop (NATW'19) and finds both stuck-at and delay fault coverage improves under pseudo-random tests using inversion TPs, and extended data collection finds noteworthy trends on the effectiveness of TP architectures."

Introduction

"Many methods improve RPR fault coverage and reduce test lengths of pseudo-random test patterns, with a common technique being test point (TP) insertion (TPI) [41]."

"Other methods of improving pseudo-random test effectiveness include changing the nature of random stimuli, such as deterministic seeding [42] and pattern weighting [43], but modifying circuit logic during test using control and observe TPs is still widely used due to their ease-of-implementation on circuit netlists."

"This article extends work presented at the North Atlantic Test Workshop (NATW'19) [44]; it was the first article to explore the effectiveness of inversion TPs at improving stuck-at and delay fault coverages compared to conventional TP architectures."

"Experiments empirically demonstrate inversion TPs frequently improve stuck-at fault coverage compared to conventional control TPs."

"Experiments demonstrate inversion TPs frequently improve delay fault coverage compared to conventional control TPs while not degrading stuck-at fault coverage."

Motivation

"ATE applies (and observes) tests at limited speeds due to the latency of external circuit connections: this impacts test quality since tests must be applied at a circuit's designed-speed (found through delay simulation or static timing analysis) to detect delay-causing defects [45, 46]."

"TPs modify circuits during test (but do not change the circuit function outside of test) and make RPR faults easier to excite and observe."

"Control TPs attempt to make the excitation and observation of faults more likely under random stimulus by forcing logic values during test."

"Signal-controlling TPs use an extra test enable (TE) pin (or scannable register [47]) to force logic in a circuit to a controlled value."

"Forcing circuit lines during test has two effects: (1) RPR faults can be easier to excite under random stimulus, and (2) RPR faults can be propagated to circuit outputs by activating circuit paths."

Inversion-Based TP Architecture

"The advantages of inversion TPs over conventional control TPs are (1) their ability to excite both stuck-at 0 and stuck-at 1 faults when activated and (2) their ability to allow stuck-at faults on inverted lines to propagate through TPs."

"Since inversion TPs do not force circuit lines to set values, they allow transitions on TP outputs, and these transitions can excite delay faults on the TP output (or on logic driven by the TP)."

"Using XORs as TPs was allowed in [48–50], but these studies did not compare inversion TPs against conventional control TPs, nor did they analyze TPs' effect on delay fault coverage."

"Previous studies noted inversion TPs may not be allowable if circuit performance is a concern to circuit designers [49]: this is because the XOR gates which implement inversions will add more delay compared to the AND/OR gate counterparts of control-0/1 TPs."

Experimental Setup

"Tsai and others [51] is chosen because it can insert any TP type without favoring one architecture over another: the algorithm does not require specific information on the TPs used and instead only knows the calculated effect a circuit modification will have on fault coverage."

"In this algorithm, the controllability and observability program (COP) [52] calculates fault coverage using probabilistic metrics."

"The TPI algorithm iteratively calculates the impact each candidate TP will have on fault coverage by re-calculating controllability and observability values when the TP is active."

"The TPI program then inserts the TP with the highest positive impact on the calculated fault coverage."

"Although other TPI from literature can insert inversion TPs, it is presumed that the TPI algorithm used will not impact this study's conclusions if the TPI method does not give an advantage to any TP architecture."

Results and Discussion

"A second observation from fault coverage results, which complements results of the original study [44], is although the delay fault coverage obtained by the various TPI methods increases, inversion TPs most frequently increase delay fault coverage the most."

"The original study [44] concluded that delay fault coverage decreases when control TPs are present: this was found by comparing the TP combinations "I/O" and "C/O"."

"When observe TPs are allowed by themselves ("O"), they rarely obtain the highest fault coverage (only for c2670 and c7553, with the latter only for delay fault coverage)."

"Allowing observe TPs to "complement" control/inversion TPs infrequently increases fault coverage, but instead more frequently decreases it."

"Allowing more TPs to decrease fault coverage is a counterintuitive result, but an explanation exists: observe TPs can only increase observability, but control and inversion TPs can increase both controllability and observability."

Conclusion and Future Directions

"This article presented the impacts of different TP architectures on stuck-at and delay fault coverage, most notably inversion TPs."

"Results showed inversion TPs frequently increase stuck-at fault coverage more than conventional TP alternatives while simultaneously increasing delay fault coverage."

"This is best shown through the benchmark c1908: control TPs only ("C") provided delay and stuck-at fault coverage significantly higher than any other TP architecture."

"Observe and control TPs must be a focus of future studies, as their true impact on delay fault detection is not yet known."

"Delay faults were modeled as TDFs due to their ease of implementation and their ability to model known defects (i.e., stuck-open faults) [53], but if the fault model includes the magnitude of a delay (as is the case of small delay defects [54]), the presence of observe and control TPs may decrease delay fault coverage."

"It may be possible to insert inversion TPs outside of critical timing paths while still obtaining favorable fault coverage, but this will be explored with future empirical data."

Single Stuck-at-Faults Detection Using Test Generation Vector and Deep Stacked-Sparse-Autoencoder [55]

This is a machine-generated summary of:

Malihi, Leila; Malihi, Razieh: Single stuck-at-faults detection using test generation vector and deep stacked-sparse-autoencoder [55]

Published in: SN Applied Sciences (2020)

Link to original: https://doi.org/10.1007/s42452-020-03460-0

If you want to cite the papers, please refer to the original.

For technical reasons we could not place the page where the original quote is coming from.

Abstract-Summary

"This paper proposed a new method for testing digital circuits without hardware implementation."

"This data-based method detects hundreds of single stuck-at faults in the ALU circuits, utilizing deep stacked-sparse-autoencoder (SSAE)."

"ATALANTA software is one of the free automatic test pattern generation tools which cover faults in high accuracy."

"Test vectors which are extracted from bench circuits via ATALANTA software are the key point of the paper."

"SSAE network is trained using the test vectors."

"Network performance is tested by changing sparse coefficients, number of stacked autoencoder and data augmentation."

"The results obtained from the deep neural network show its significant performance in the circuit faults detection automatically."

Introduction

"There are some basis automatic test pattern generation (ATPG) algorithms, such as D-algorithm [56], path oriented decision making (PODEM) [57] and FAN [58, 59] to distinguish stuck-at faults."

"Agrawal and others [60] distinguish a part of faults with generating test vectors and for the remaining faults usages branch and bound technique."

"In the last century, the use of artificial intelligence methods feature extraction and fault detection has expanded."

"These methods extract feature from the output voltage and input signal of the circuit and compare it with the target to detect the faults."

"Fault detection using support vector machine is purposed in [61, 62]."

"One of the newest methods in the field of automatic feature extraction and fault detection is deep learning."

"Testing effort is utilized in this method since it has a high effect on fault detection and correction process."

"The purpose of the paper is to identify the single stack-at faults based on the automatic test patterns generation (ATPG)."

Materials and Methods

"The ALU unit is the last component to perform computation and processing on the data."

"This benchmark performs binary and BCD arithmetic, and besides logic and shift operations."

"ATALANTA programming environment is utilized to do coding test pattern generator for stuck-at fault detection in combinational circuits."

"Test vectors for both of stuck-at faults were generated by ATALANTA."

"If the training data is very little, image augmentation will require to increase the accuracy of deep networks."

"Image augmentation artificially can be done by random rotation, shifts, shear and flips, etc Autoencoder is a specific type of artificial neural network used to encode learning, and then uncompressed that code into something which closely matches the original data."

"Of training the network and predicting the target value of Y, in return for the X input, an autoencoder will be trained to rebuild the X input [63]."

Sparse Auto-Encoder

"By applying sparse limitations on hidden units, autoencoder detects fascinating structures in the data, even if the number of hidden layers is high."

"The main components of SSAE are many sparse auto-encoders (SAE) and a softmax classifier."

"Of all, each SAE and the softmax layer are trained."

"More details about the layers are as follows: The main layers of an auto-encoder (AE) are the input layer, hidden layer and reconstruction layer, respectively [63]."

"The input data is encoded using a hidden layer, and then reconstruction is done."

"The output of the first SAE is the second SAE input, and the training process is like the first one."

Experimental Results

"After input and output describing, the logical model of ALU circuit has been defined."

"55,752 test vectors were extracted from 8 bit ALU totally."

"The minimum fault coverage is 94.06% for 12 bit ALU and the maximum fault coverage is 98.78% for 9 bit ALU."

"The number of classes in 8 bit ALU are 3288, in 9 bit ALU are 5285 and in 12 bit ALU are 2584."

"There are 3288 faults in 8 bit ALU, hence the number of output neurons are 3288."

"For the 8 bit ALU, the validation plot (blue) is higher than others."

"The best performance is 97.5% accuracy for 8 bit ALU."

"For 9 bit ALU, the validation performance is higher."

"The results of 8 bit ALU are almost the same in both methods."

Conclusion

"Fault detection accuracy was 99.6% for 12 bit ALU."

"By comparing the proposed method with MLP, the experimental results show that the proposed method achieves high fault detection accuracy."

"The best accuracy for fault detection is related to 12 bit ALU which is 99.6%."

"Using test vectors as features due to high accuracy for fault detection."

An Efficient VLSI Test Data Compression Scheme for Circular Scan Architecture Based on Modified Ant Colony Meta-Heuristic [64]

This is a machine-generated summary of:

Mitra, Sanjoy; Das, Debaprasad: An Efficient VLSI Test Data Compression Scheme for Circular Scan Architecture Based on Modified Ant Colony Meta-heuristic [64]

Published in: Journal of Electronic Testing (2020)

Link to original: https://doi.org/10.1007/s10836-020-05880-7

If you want to cite the papers, please refer to the original.

For technical reasons we could not place the page where the original quote is coming from.

Abstract-Summary

"A stochastic heuristic based bio-inspired optimization approach namely ant colony algorithm (ACO) is applied after modification and customization to improve compression efficiency."

"In circular scan architecture, test data compression is achieved by updating the conflicting bits between the most recently captured response and test vector to be applied next."

"The quantity of conflicting bits also manifests the Hamming distance between the most recently captured response and the next test vector."

"A significant reduction in test data volume and test application time is achieved by reducing Hamming distance."

"The test vectors are presumed as cities and Hamming distance between a pair of test vectors is treated as intercity distance and a modified ACO algorithm in combination with mutation operator is applied here to resolve this combinatorial optimization problem."

"An average improvement of 6.36% in compression ratio and 4.77% in test application time is achieved."

Introduction

"Test-data compression [65] offers a promising solution to the problem of reducing the test data volume for SoCs and to reduce the test time, peak power and average power."

"An alternate approach for reducing test data volume for SoCs is based on the use of data compression techniques such as statistical coding, run-length coding and Golomb coding [65–67]."

"The key point of determining the test data compression efficiency of circular scan based testing is minimizing the number of conflicting bits between the template pattern and the real test pattern."

"To reduce test data volume, this paper proposes a new test vector reordering scheme which can be equivalently thought off as an optimization of corresponding traveling salesman problem (TSP) and this is NP – complete by nature."

"This section also briefs about the proposed meta-heuristic of Modified and Customized Ant Colony Optimization (MCACO) for solving the traveling salesman problem renovated from conflicting bit minimization problem of circular scan test applications."

Review of Circular Scan Test Application and Role of Conflicting Bits in Test Data Compression

"The bits of a test pattern, corresponding to the scan cells that have equal distance to the scan inputs, constitute a test slice."

"In the current test slice, if a set of bits differs from the corresponding bits in the scan cells, one of them is selected through the decoder and the value in the scan cell is updated through the data input."

"Since only one of the conflicting bits of the current test slice can be changed at each cycle, multiple full rotations of the scan chain content may be required."

"The test slice with the highest number of conflicting bits of a test pattern determines how many times the scan chains have to be fully rotated to apply the current pattern."

"The inverting path is selected for scan chain that has the conflicting bits."

Formulation of Conflicting Bit Minimization (CBM) as an Optimization Problem

"A conflicting bit minimization problem is formulated conceding three scan chains with eight scan cells each are contained in the example circuit under test (CUT)."

"To achieve the test mode of circular scan, in that the previous response is used as the next patterns' template, the conflicting bits between them should be updated."

"The number of conflicting bits between R_3 and V_2 are computed as 8 bits which is also the Hamming distance between them."

"Different orders of the test patterns will result in a different number of conflicting bits."

"The order of V_3, V_2, V_5, V_1, and V_4 has the number of conflicting bits $W_{32} + W_{25} + W_{51} + W_{14} = 11 + 14 + 12 + 11 = 48$, where W_{ij} denotes the weight of the two vertexes of i and j. The order of the test patterns influences the number of conflicting bits that need to be updated, and that number should be minimized."

Traveling Salesman Problem (TSP) and Conflicting Bit Minimization (CBM) Analogy

"To solve the problem of the conflicting bits minimization problem, which is NP-complete, the modified and customized ant colony optimization algorithm (MCACO) is proposed here in this paper."

"In the proposed MCACO algorithm, the global optimum distance from a set of feasible distances traveled by a crowd of ants correspond to an optimistic solution to the conflicting bit minimization problem and fitness of the solution resembles the total tour distance covering all the cities."

"In the projected MCACO, this pheromone amount is defined and calculated by where R is the random number between the limits [0,1] and $L^m(i)$ is the length of the tour of the best ant."

"Pheromone matrix (σ) and distance matrix (h) are initialized at the beginning of MCACO alike other ACO based TSP solutions given the specific problem to be solved."

"In ACO, the intensity of pheromone determines the path to be followed by the ant for search of food particles."

Application of Ant Colony Algorithm for Conflicting Bit Minimization

"In previous sections, given test vectors renovated as cities in TSP are subjected to MCACO algorithm for computing an optimal order of test vectors so that the extent of conflicting bits between them gets minimized."

"In the proposed MCACO algorithm, the global optimum distance from a set of expedient distances traveled by a crowd of ants corresponds to an optimal solution to the reordering problem and fitness of the solution resembles the total tour distance covering all the cities."

"Reinstate individual ants to its initial position and compute the least magnitude against each ant-to-ant value and these values are to be stored in a vector."

"The fitness value estimation corresponding to each ant is given by The most suitable value of the ant denotes the optimal sequence of the test vector."

Testing Setup, Experimental Results and Analysis

"Against scan chain length of 256, i.e., SC#256, an improvement of 9.79% in the average reduction in TAT is recorded corresponding to the benchmark circuit s13207 and an improvement of 1.69% in TAT reduction is seen in comparison to the very recent circular scan test based compression method in [68]."

"An average improvement in TAT reduction of 21.83% is noticed in SC#64 corresponding to the benchmark circuit s38584 and an improvement of 2.03% is observed in comparison with circular scan test based compression method in [68]."

"In case of the #SC = 128, significant improvement (22.05%) in average reduction % of test application time (TAT) is observed corresponding to the benchmark circuit s15850 and 3% improvement in TAT reduction is observed when we compare with the work in [68]."

Conclusion

"The volume of such conflicting bits is minimized in order to get better compression ratio and test application time."

"Experimental results justify our claim of notable improvement in compression ratio and test application time (TAT) reduction over various recent methods."

"We have only modified the pheromone update meta-heuristic of ant colony optimization technique leaving behind the scope of modifying the other slants of ant colony method and this may even yield further improved results in terms of compression ratio and test application time reduction in circular scan based test applications."

"Even if a guided mutation is applied in place of random mutation, there is a probability of further 2% to 3% average improvement in compression ratio and test application time."

On Reducing Test Data Volume for Circular Scan Architecture Using Modified Shuffled Shepherd Optimization [69]

This is a machine-generated summary of:

Jayabalan, Muralidharan; Srinivas, E.; Shajin, Francis H.; Rajesh, P.: On Reducing Test Data Volume for Circular Scan Architecture Using Modified Shuffled Shepherd Optimization [69]

Published in: Journal of Electronic Testing (2021)

Link to original: https://doi.org/10.1007/s10836-021-05975-9

If you want to cite the papers, please refer to the original.

For technical reasons we could not place the page where the original quote is coming from.

Abstract-Summary

"A novel test data compression (TDC) system is proposes for reducing a test data volume (TDV) for circular scan (CS) architecture."

"A modified version of meta-heuristic population based optimization approach, hence it is called shuffled shepherd optimization (SSO) and it is used to optimize the conflicting bit minimization (CBM) problem."

"Reduced test data volume and test application time are achieved by diminishing the Hamming distance between the currently captured response and the subsequent test vector."

"Experimental outcomes display the proposed TDC method using MSSO achieves an average development of 8.36% in compression ratio (CR) as well as 6.77% in TAT is attained by reducing a TDV."

"This proposed approach outperforms other state-of-art test data compression schemes by providing improved CR, TAT reduction and TDV reduction."

Introduction

"Test data compression schemes [70] act as a hopeful solution for solving the issue of TDV reduction to SoCs."

"Some recent research related to the presented work is reviewed in this section: Thilagavathi and others [71] have presented a new test power and volume reduction technique by integrating hybrid X-fill and two-stage test data compression (TS-TDC) schemes to use in digital VLSI circuits."

"Chakrapani and others [72] have presented an effective method of compression of test data depending on stored set named Lingering Component Reduction (LCR) Code."

"Vohra and others [70] have presented an efficient test data compression scheme that maximizes test data transport time and tester memory."

"A test data is dividing into blocks of standard sizes that are compressed utilizing2-stage encoding method."

"Test-data compression provides a promising solution to issues of minimizing TDV to SoCs, to reducing a test time, peak power, average power."

Preliminaries

"One of the effective designs for testing (DFT) architecture is Circular Scan (CS)."

"For reduce the TDV, the determined test patterns are loaded with the help of the one-bit update and transmission mode designs."

"Reordering a set of test vectors can be considered a travelling salesman problem (TSP)."

"Hamming distance based reordering scheme aids for rearranging the test vectors both in vertical and horizontal directions."

"Reordering a set of test vectors based on the Hamming distance can be used with an efficient compression technique to yield the maximum compression ratio."

"Test vector reordering techniques reduces the switching function of the CUT by reducing in switching function on the inputs during the circuit testing process."

"Those heuristic approaches find the shortest path, which helps in optimizing the problem of test vector reordering."

Proposed Methodology

"Another advantage of the circular scan design is that captured response from the old test pattern is utilized to the template for novel test pattern and conflicting bits of the new test pattern need to be updated."

"The modified optimization algorithm minimizes the number of conflicting bits among response of one test vector, subsequent test vector."

"A minimization of 22 conflicting bits is achieved with the help of modified optimization algorithm."

"In this MSSO is used to select the optimal shortest path to find an amount of conflicting bits is reduced by selecting optimally selecting the shortest path between the test vectors."

"The use of mutation in the proposed modified approach helps in further reducing the number of conflicting bits via reordering the test vector sets."

"After performing a fixed number of repetitions, this algorithm reaches the stop condition by obtaining an optimal order of the path, which reduces that number of conflicting bits needed to solve the CBM problem."

Experimental Results and Analysis

"From TAT reduction of #SC = 64 shows the proposed strategy reaches 21.4% and 18% higher TAT reduction than the two existing approaches corresponding to benchmark circuit s13207."

"From TAT reduction of #SC = 128 shows the proposed strategy reaches 45% and 24.2% higher compression ratio than the two existing approaches corresponding to benchmark circuit s13207."

"From TAT reduction of #SC = 256 shows the proposed strategy reaches 50% and 6% higher TAT reduction than the two existing approaches corresponding to benchmark circuit S13207."

"Test data volume of #SC = 64 shows the proposed strategy reaches 37% and 77% higher TDV reduction than the two existing approaches corresponding to benchmark circuit s13207."

"From TDV reduction of #SC = 128 shows the proposed strategy reaches 14.2% and 7% higher TDV reduction than the two existing approaches corresponding to benchmark circuit s13207."

Conclusion

"A novel TDC scheme is proposed for reduce the TDV to CS architecture."

"In CS architecture, TDC is reached by upgrading only the conflict bits among template pattern, real test pattern."

"Experimental outcomes display the proposed TDC algorithm using MSSO achieves an average 8.36% development on compression ratio, 6.77% reduction on test application time, and 5.68% in reducing the test data volume."

"The proposed approach outperforms other state-of-art test data compression schemes by providing improved compression ratio, TAT reduction and TDV reduction."

An Efficient Partitioning and Placement Based Fault TSV Detection in 3D-IC Using Deep Learning Approach [73]

This is a machine-generated summary of:

Radhakrishnan Nair, Radeep Krishna; Pothiraj, Sivakumar; Radhakrishnan Nair, T. R.; Cengiz, Korhan: An efficient partitioning and placement based fault TSV detection in 3D-IC using deep learning approach [73]

Published in: Journal of Ambient Intelligence and Humanized Computing (2021)

Link to original: https://doi.org/10.1007/s12652-021-02964-w

Copyright of the summarized publication:

If you want to cite the papers, please refer to the original.

For technical reasons we could not place the page where the original quote is coming from.

Abstract-Summary

"Fault Through Silicon Via (TSV) detection is a bottleneck, due to poor fabrication processes such as partitioning and placement."

"To resolve these shortcomings, this paper proposes an Efficient Partitioning and Placement based Fault TSV detection in 3D-IC."

"The proposed work comprises five processes: Quick cut oriented Partitioning, Multi-Objective based Placement, Deep learning based Fault TSV detection, Re-routing and Adaptive Time Division Multiple Access (TDMA) time slot."

"Quick Cut algorithm has been employed to partition the 3D-IC and it is easier for placement process."

"The placement is executed through Multi-Objective Brain Storm Optimization algorithm that selects the optimal place to position the cells in 3D-IC."

"The fault TSV in the 3D-IC is detected using the Adam Deep Neural Network algorithm."

"The Adaptive TDMA algorithm has been used to provide time slot to TSV positioned in each partition."

Introduction

"The fabrication of 3D-IC has been achieved through partitioning, placement and TSV [74]."

"These partitioning techniques have not concentrated on the area constraints of the 3D-IC and hence, difficulties are created to the placement process."

"The placement is the next process to the partitioning in 3D-IC fabrication [75]. The effective placement process in 3D-IC fabrication reduces the wirelength and area size drastically."

"These placement techniques consume more time and induce congestion to the 3D-IC."

"The state-of-art partitioning techniques cause congestion during the placement of circuits, due to the absence of significant parameter consideration (size of the circuit, TSV count)."

"These issues are overwhelmed by the proposed partitioning and placement based fault TSV detection using deep learning technique."

"The proposed method reduces the wirelength and area of the 3D-IC using two significant processes: (i) QC oriented partitioning and (ii) Multi-Objective Brain Storm Optimization (MO-BSO) based placement."

State-of-Art

"Further, min cut partitioning algorithm is implemented to partition the circuits in the 3D-IC."

"Banerjee and others [76] have introduced graph based partitioning method in 3D-IC fabrication."

"Tuan and others have pointed out the hybrid optimization based placement in the 3D-IC circuit [77]."

"The fault TSV detection in 3D-IC increases the fabrication performance without any defects."

"Huang and others [78] have proposed machine learning approach to detect the defects in the TSV of 3D-ICs."

"The physical parameters of the TSV are measured to detect the defects in 3D-IC."

"Based on the measured electrical characteristics of the 3D-IC, the fault present in the TSV is detected."

"The limitations of the above-mentioned state-of-art works are resolved by the proposed "Partitioning and Placement based Fault TSV detection in 3D-IC by using the Deep Learning approach"."

Problem Statement

"The major problem encountered from the preceding works associated with partitioning, placement and fault TSV detection are listed as follows, 1."

"Absence of re-routing the signal transmitted through the defected TSV.As a result, the delay is increased in the signal transmission of 3D-ICs."

"Time Driven Placement (TDP) is presented for the 3D-IC placement process and it follows the global placement procedure [79]."

"Integer Linear Programming (ILP) model based fault tolerance technique, which detects defect TSV and allocates spare TSV based on chip yields, has been used Sait and others [80]."

"ILP based 3D-IC cannot achieve effective fault tolerance in defect detection."

"Time Division Multiplexing Access (TDMA) based Fault Tolerance (TFT) technique has been offered to detect the fault present in the TSV where delay during signal transmission is considered [81]."

Proposed Work

"The proposed QC algorithm partitions the circuit based on the size of the circuit, TSV count and Input/Output (I/O) bin count."

"The proposed algorithm starts with initializing the population and it clusters the solutions randomly with the size S. The weight factor for partitions in each cluster is estimated to find the best solution."

"The weight factor is estimated using five sequential parameters including cross-talk, temperature, wirelength, power and size of the partitions and these metrics are used to reduce the wirelength and enhance the effective signal transmission in 3D-IC."

"From the high weight TSV, the nearby defect free TSV is selected to transmit the signal to the other circuits that exist in the 3D-IC."

"The proposed work provides an adaptive time slot to the TSV present in each partition in order to reduce the delay during the signal transmission between different dies."

Experimental Study

"In this simulator, the proposed algorithms are run for implementing partitioning, placement and fault TSV detection in 3D-IC fabrication."

"The proposed method obtains a reduced area in 3D-IC fabrication compared to the existing methods and it is acquired through the proposed MO-BSO based placement process."

"The proposed method reduces the area of the 3D-IC up to 30% compared to the existing Tabu Search and TSV Fault Tolerance methods."

"From the comparison results, it is seen that the proposed method achieves less running time compared to the existing methods, since the proposed 3D-IC fabrication utilizes the effective algorithm for partitioning, placement and fault TSV detection."

"Comparison, it is concluded that the proposed mechanism achieves better results in 3D-IC fabrication compared to the existing Tabu Search and TSV Fault Tolerance methods."

Conclusion

"An effective fabrication mechanism has been proposed for 3D-IC using the partitioning, placement and fault TSV detection."

"The partitioning has been performed using the Quick Cut algorithm which has taken into account the size of circuit, I/O bins and TSV count parameters."

"The fault TSV detection is executed using the ADNN algorithm which considers resistance, TSV stress, Signal Transmission Time and Voltage metrics."

"The Adaptive TDMA algorithm is used and it considers CSR, the number of defect TSV and number of suspicious TSV to allocate slots adaptively."

References

1. Farooq U, Mehrez H (2021) Pre-silicon verification using multi-FPGA platforms: a review. J Electron Testing. https://doi.org/10.1007/s10836-021-05929-1
2. Kern C, Greenstreet MR (1999) Formal verification in hardware design: a survey. ACM Trans Des Automat Electron Syst 4(2):123–193
3. Patil S, Scholten S, Tao M, Al-Asaad H (2019) Survey of memory, timing, and power management verification methods for multi-core processors. In: 2019 IEEE 10th Annual Information Technology, Electronics and Mobile Communication Conference (IEMCON). IEEE, pp 0110–0119
4. M. Graphics (2020) https://www.mentor.com/products/fv/modelsim/
5. Farooq U, Baig I, Alzahrani BA (2018) An efficient inter-FPGA routing exploration environment for multi-FPGA systems. IEEE Access 6:56301–56310
6. Tang Q, Tuna M, Mehrez H (2014) Performance comparison between multi-FPGA prototyping platforms: hardwired off-the-shelf, cabling, and custom. In: 2014 IEEE 22nd Annual International Symposium on Field-Programmable Custom Computing Machines. IEEE, pp 125–132
7. Kumar B, Fujita M, Singh V (2019) SAT-based silicon debug of electrical errors under restricted observability enhancement. J Electron Testing. https://doi.org/10.1007/s10836-019-05830-y
8. Kumar B, Basu K, Fujita M, Singh V (2020) Post-silicon gate-level error localization with effective and combined trace signal selection. IEEE Trans Comput Aided Des Integr Circ Syst 39:248–261
9. Vali A, Nicolici N (2018) Bit-flip detection-driven selection of trace signals. IEEE Trans Comput Aided Des Integr Circ Syst 37(5):1076–1089
10. Moness M, Gaber L, Hussein AI, Ali HM (2022) Automated design error debugging of digital VLSI circuits. J Electron Testing. https://doi.org/10.1007/s10836-022-06020-z
11. Gaber L, Hussein AI, Mahmoud H, Mabrook MM, Moness M (2020) Computation of minimal unsatisfiable subformulas for SAT-based digital circuit error diagnosis. J Ambient Intell Human Comput 13:3693–3711
12. Jutman A, Ubar R (2000) Design error diagnosis in digital circuits with stuck-at fault model. Microelectron Reliabil 40:307–320
13. Sun Y, Millican SK (2022) Applying artificial neural networks to logic built-in self-test: improving test point insertion. J Electron Testing. https://doi.org/10.1007/s10836-022-06016-9
14. Seshan K (2018) Reliability issues: reliability imposed limits to scaling. In: Seshan K, Schepis D (eds) Handbook of thin film deposition, 4th edn. William Andrew Publishing, pp 43–62
15. Acero C, Feltham D, Liu Y, Moghaddam E, Mukherjee N, Patyra M, Rajski J, Reddy SM, Tyszer J, Zawada J (2017) Embedded Deterministic Test Points. IEEE Trans Very Large Scale Integr Syst 25(10):2949–2961

16. Ma Y, Ren H, Khailany B, Sikka H, Luo L, Natarajan K, Yu B (2019) High performance graph convolutional networks with applications in testability analysis. In: Proceedings of the 56th Annual Design Automation Conference (DAC), New York, NY, USA, vol 18. ACM, pp 1–6
17. Roy S, Stiene B, Millican SK, Agrawal VD (2019) Improved random pattern delay fault coverage using inversion test points. In: Proceedings of the IEEE 28th North Atlantic Test Workshop (NATW), Burlington, VT, USA. IEEE, pp 1–6
18. Sun Y, Millican SK (2019) Test point insertion using artificial neural networks. In: Proceedings of the IEEE Computer Society Annual Symposium on VLSI (ISVLSI), Miami, FL, USA. IEEE, pp 253–258
19. Tsai H-C, Cheng K-T, Lin C-J, Bhawmik S (1997) A hybrid algorithm for test point selection for scan-based BIST. In: Proceeding of the 34th ACM/IEEE Design Automation Conference (DAC), Anaheim, CA, USA. IEEE, pp 478–483
20. Brglez F (1984) On testability analysis of combinational networks. In: Proceedings—IEEE International Symposium on Circuits and Systems (ISCAS), Montreal, Quebec, Canada, vol 1. IEEE, pp 221–225
21. Karystinos GN, Pados DA (2000) On overfitting, generalization, and randomly expanded training sets. IEEE Trans Neural Netw 11(5):1050–1057
22. Moghaddam E, Mukherjee N, Rajski J, Solecki J, Tyszer J, Zawada J (2019) Logic BIST with capture-per-clock hybrid test points. IEEE Trans Comput Aided Des Integr Circ Syst 38(6):1028–1041
23. Millican SK, Sun Y, Roy S, Agrawal VD (2019) Applying neural networks to delay fault testing: test point insertion and random circuit training. In: Proceedings of the 28th Asian Test Symposium (ATS), Kolkata, India. IEEE, pp 13–18
24. Champac V, Mesalles J, Villacorta H, Vargas F (2021) Analysis and detection of open-gate defects in redundant structures of a FinFET SRAM cell. J Electron Testing. https://doi.org/10.1007/s10836-021-05955-z
25. Bhoj AN, Simsir MO, Jha NK (2012) Fault models for logic circuits in the multigate era. IEEE Trans Nanotechnol 11(1):182–193
26. Liu Y, Xu Q (2012) IEEE International Test Conference. IEEE
27. Harutyunyan G, Tshagharyan G, Vardanian V, Zorian Y (2014) IEEE 32nd VLSI Test Symposium (VTS). IEEE
28. Harutyunyan G, Tshagharyan G, Zorian Y (2015) Test and repair methodology for FinFET-based memories. IEEE Trans Device Mater Reliabil 15(1):3–9
29. Harutyunyan G, Tshagharyan G, Zorian Y (2015) Impact of parameter variations on FinFET faults. In: IEEE 33rd VLSI Test Symposium (VTS). IEEE
30. Chen MC, Wu TH, Wu CW (2018) A built-in self-test scheme for detecting defects in FinFET-based sram circuit. In: IEEE 27th Asian Test Symposium. IEEE, pp 19–24
31. Lin Y, Yu B, Xu B, Pan DZ (2017) Triple patterning aware detailed placement toward zero cross-row middle-of-line conflict. IEEE Trans Comput Aided Des Integr Circ Syst 36(7):1140–1152
32. Hamdioui S, Taouil M, Haron NZ (2015) Testing open defects in memristor-based memories. IEEE Trans Comput 64(1):247–259
33. Handique M, Deka JK, Biswas S (2020) An efficient test set construction scheme for multiple missing-gate faults in reversible circuits. J Electron Testing. https://doi.org/10.1007/s10836-020-05855-8
34. Rahaman H, Kole DK, Das DK, Bhattacharya BB (2008) On the detection of missing-gate faults in reversible circuits by a universal test set. In: Proceedings of 21st international conference on VLSI design, VLSID. IEEE, pp 163–168
35. Rahaman H, Kole DK, Das DK, Bhattacharya BB (2011) Fault diagnosis in reversible circuits under missing-gate fault model. Comput Electr Eng 37(4):475–485

36. Zhang H, Wille R, Drechsler R (2010) SAT-based ATPG for reversible circuits. In: Proceedings of 5th international design and test workshop. IEEE, pp 149–154
37. Surhonne AP, Chattopadhyay A, Wille R (2017) Automatic test pattern generation for multiple missing gate faults in reversible circuits. In: Proceeding of international conference on reversible computation. Springer, Cham, pp 176–182
38. Kole A, Wille R, Datta K, Sengupta I (2017) Test pattern generation effort evaluation of reversible circuits. In: Proceedings of international conference on reversible computation. Springer, Cham, pp 162–175
39. Fooladi M, Kamran A (2020) Speed-up in test methods using probabilistic merit indicators. J Electron Testing. https://doi.org/10.1007/s10836-020-05871-8
40. Roy S, Stiene B, Millican SK, Agrawal VD (2020) Improved pseudo-random fault coverage through inversions: a study on test point architectures. J Electron Testing. https://doi.org/10.1007/s10836-020-05859-4
41. Hayes JP, Friedman AD (1974) Test point placement to simplify fault detection. IEEE Trans Comput C-23(7):727–735
42. Xiang D, Wen X, Wang L-T (2017) Low-power scan-based built-in self-test based on weighted pseudorandom test pattern generation and reseeding. IEEE Trans Very Large Scale Integr Syst 25(3):942–953
43. Rajski J, Tyszer J (1998) Arithmetic built-in self-test for embedded systems. Prentice-Hall, Upper Saddle River
44. Roy S, Stiene B, Millican SK, Agrawal VD (2019) Improved random pattern delay fault coverage using inversion test points. In: IEEE 28th North Atlantic Test Workshop (NATW). IEEE, pp 206–211
45. Majhi AK, Agrawal VD (1998) Delay fault models and coverage. In: Proceedings of 11th international conference on VLSI design, Chennai, India. IEEE, pp 364–369
46. Nigh P, Needham W, Butler K, Maxwell P, Aitken R (1997) An experimental study comparing the relative effectiveness of functional, scan, IDDq and delay-fault testing. In: Proceedings of 15th IEEE VLSI Test Symposium. IEEE, pp 459–464
47. Yang J, Touba NA, Nadeau-Dostie B (2012) Test point insertion with control points driven by existing functional flip-flops. IEEE Trans Comput 61(10):1473–1483
48. Fang Y, Albicki A (1995) Efficient testability enhancement for combinational circuit. In: Proceedings of international conference on computer design (ICCD). IEEE, pp 168–172
49. Ren H, Kusko M, Kravets V, Yaari R (2009) Low cost test point insertion without using extra registers for high performance design. In: Proceedings of the International Test Conference (ITC), Austin, TX. IEEE
50. Rudnick EM, Chickermane V, Patel JH (1994) An observability enhancement approach for improved testability and at-speed test. IEEE Trans Comput Aided Des Integr Circ Syst 13(8):1051–1056
51. Tsai HC, Cheng K-T, Lin CJ, Bhawmik S (1997) A hybrid algorithm for test point selection for scan-based BIST. In: Proceedings of the 34th design automation conference (DAC). IEEE, pp 478–483
52. Brglez F (1984) On testability analysis of combinational networks. In: Proceedings of the international symposium on circuits and systems (ISCAS), vol 1. IEEE, pp 221–225
53. Mahmod J, Millican SK, Guin U, Agrawal VD (2019) Special session: delay fault testing—present and future. In: Proceedings of the 37th VLSI Test Symposium (VTS), Monterey, CA. IEEE
54. Mattiuzzo R, Appello D, Allsup C (2009) Small-delay-defect testing. EDN 54(13):28
55. Malihi L, Malihi R (2020) Single stuck-at-faults detection using test generation vector and deep stacked-sparse-autoencoder. SN Appl Sci. https://doi.org/10.1007/s42452-020-03460-0
56. Roth JP et al (1968) Diagnosis of automata failures. IBM J Res Dev 10(4):278–291
57. Goel P (1981) An implicit enumeration algorithm to generate tests for combinational logic circuits. IEEE Trans Comput C-30(3):215–222. https://doi.org/10.1109/TC.1981.1675757

58. Fujiwara H, Shimono T (1983) On the acceleration of test generation algorithms. IEEE Trans Comput C-32(12):1137–1144
59. Fujiwara H (1985) Fan: a fanout-oriented test pattern generation algorithm. In: Proceedings of IEEE international symposium circuits system. IEEE, pp 671–674
60. Agrawal A, Saldanha A, Lavagno L, Sangiovanni AL (1997) Compact and complete test set generation for multiplestuck-faults. In: Proceedings of IEEE/ACM ICCAD, San Jose, CA, USA. IEEE, pp 212–219
61. Qing HU, Wang RJ, Zhan YJ (2008) Fault diagnosis technology based on SVM in power electronics circuit. Proc Chin Soc Electr Eng 28(12):107–111
62. Cui J, Wang Y, Liu Q (2007) The technique of power electronic circuit fault diagnosis based on higher-order spectrum analysis and support vector machines. Proc Chin Soc Electr Eng 10(10):62–66
63. Xu J, Xiang L, Liu Q et al (2016) Stacked sparse autoencoder (SSAE) for nuclei detection of breast cancer histopatholog images. IEEE Trans Med Imaging 35(1):119–130
64. Mitra S, Das D (2020) An efficient VLSI test data compression scheme for circular scan architecture based on modified ant colony meta-heuristic. J Electron Testing. https://doi.org/10.1007/s10836-020-05880-7
65. Chandra A, Chakrabarty K (2003) Test data compression and test resource partitioning for system-on-a-chip using frequency-directed run-length (FDR) codes. IEEE Trans Comput 52(8):1076–1088
66. Ruan X, Katti RS (2006) Data-independent pattern run-length compression for testing embedded cores in SoCs. IEEE Trans Comput 56(4):545–556
67. El-Maleh AH (2008) Test data compression for system-on-a-chip using extended frequency-directed run-length code. IET Comput Digit Tech 2(3):155–163
68. Zhang L, Mei J, Yan B (2018) New test set compression scheme for circular scan. EURASIP J Embed Syst 1:1
69. Jayabalan M, Srinivas E, Shajin FH, Rajesh P (2021) On reducing test data volume for circular scan architecture using modified shuffled shepherd optimization. J Electron Testing. https://doi.org/10.1007/s10836-021-05975-9
70. Vohra H, Singh A, Gill SS (2020) An innovative two-stage data compression scheme using adaptive block merging technique. Integration 73:68–76
71. Thilagavathi K, Sivanantham S (2018) Two-stage low power test data compression for digital VLSI circuits. Comput Electr Eng 71:309–320
72. Chakrapani K, Muthaiah R (2018) Test data compression using Lingering Component Reduction technique for system-on-a-chip applications. Comput Electr Eng 66:1–3
73. Radhakrishnan Nair RK, Pothiraj S, Radhakrishnan Nair TR, Cengiz K (2021) An efficient partitioning and placement based fault TSV detection in 3D-IC using deep learning approach. J Ambient Intell Human Comput. https://doi.org/10.1007/s12652-021-02964-w
74. Salah K (2014) TSV-based 3D integration fabrication technologies: an overview. In: 2014 9th international design and test symposium (IDT). IEEE, pp 253–256
75. Lin SE, Kim DH (2019) Wire length characteristics of multi-tier gate-level monolithic 3D ICs. IEEE Trans Emerg Topics Comput 7(2):301–310
76. Banerjee S, Majumder S, Bhattacharya BB (2014) A graph-based 3D IC partitioning technique. In: 2014 IEEE computer society annual symposium on VLSI, pp 613–618. https://doi.org/10.1109/ISVLSI.2014.82
77. To TA, Tuan DA, Thanh VC, Siddiqi UF, Shiraishi Y, Motegi K (2013) A hybrid particle swarm optimization for component placement in 3D IC design. In: 2013 IEEE electrical design of advanced packaging systems symposium (EDAPS). IEEE, pp 68–71
78. Huang YJ, Pan CL, Lin SC, Guo MH (2018) Machine-learning approach in detection and classification for defects in TSV-based 3-D IC. IEEE Trans Comp Packag Manuf Technol 8(4):699–706. https://doi.org/10.1109/TCPMT.2017.2788896

79. Reddy RP, Acharyya A, Khursheed S (2017) A cost-effective fault tolerance technique for functional TSV in 3-D ICs. IEEE Trans VLSI Syst 25(7):2071–2080. https://doi.org/10.1109/TVLSI.2017.2681703
80. Sait SM, Oughali FC, Al-Asli M (2016) Design partitioning and layer assignment for 3D integrated circuits using tabu search and simulated annealing. J Appl Res Technol 14(1):67–76. https://doi.org/10.1016/j.jart.2015.11.001
81. Wu G, Chu C (2017) Two approaches for timing-driven placement by Lagrangian relaxation. IEEE Trans Comput Aid Des Integr Circ Syst 36(12):2093–2105. https://doi.org/10.1109/TCAD.2017.2697947

Chapter 6
Optimization Approaches for Clocking and Delay Minimization

Introduction by the Editor

Clocking is another critical step in VLSI chip design. Majority of the design blocks are sequential in nature and can't work without a clock source and clock delivery structure in place. Most events which are synchronous in nature happen on clock edges. While implementing a clock tree in the design, we are faced with multiple challenges like clock synchronization, clock delay, clock skew and jitters. The following chapter discusses techniques used for clock optimizations and delay reduction of better performance. Various algorithms like performance analysis of nano-scaled FinFETs, Radiation hardened NMOS only Schmitt trigger-based latch design, a novel 4:2 compressor architecture for high performance application, clock skew reduction for peak current reduction is discussed in the chapter.

MOSFETs and FinFETs are discussed in this chapter with pros and cons of each. The data points, because of comparison, are verified using simulations with a certain power supply, channel widths and lengths. This chapter's conclusion is based on the leakage power reduction techniques being used based on usage of circuits with various channel widths and lengths. This chapter emphasizes using focused low leakage input dependent approach provides a great deal in reducing leakage. Soft error modelling of radiation hardened inverters is also discussed. NST-VB based latch designs and its comparison with other latch designs is mentioned. The study highlights that the critical charge increases with voltage supply for inverter designs. It is due to the obvious high supply voltage creating a bigger channel for current conduction which will also increase charge on critical nodes of the circuit. The critical charge on the inverter is higher than other logic components and is complimentary for latch-based design to be radiation hardened.

Transistor level design has always been important in VLSI design cycle especially in power consumption of the circuit. Quad rail, dual rail are elaborated in this chapter in the essence of removing time reference logic in place. A sample design is

A. S. Shastri (ed.), *Optimization Methods in VLSI Design*,
https://doi.org/10.1007/978-981-95-2457-0_6

implemented using numerous null conventional logics, its effectiveness is also discussed. Cuckoo optimization techniques is used for buffer insertion in VLSI circuits, this greatly helps preserving signal strength over long interconnects of clocks. The number and size of the buffer is estimated for the necessary inductance effect. Two case studies are highlighted in this chapter. One is increasing inductance per unit length; propagation delay is reduced while buffer insertion. Another one is increasing the inductance of the interconnect wire, propagation delay rate decreases with buffer insertion. This chapter concludes that the buffer insertion failure may lead to the use of wire interconnect as an effective technique.

This chapter also includes sub-threshold timing analysis in detail. Lot of experiments are discussed which are based on equalization of delay buffers for specific cases. It is also inferred that reduction in current drives can result in slow voltage scaling. Expanding on the same, at scaled voltage levels, it's important to mention that the current drive is smaller, the delay will be higher. The power dissipation in the sub-threshold region is closely linked to gate length changes. Threshold voltage and sub-threshold current greatly rises with decrease in scaled voltage. Monte Carlo analysis mentioned significant increase in delay and power dissipation with changed threshold voltage and channel length.

Machine Generated Summaries

Disclaimer: The summaries in this chapter were generated from Springer Nature publications using extractive AI auto-summarization: An extraction-based summarizer aims to identify the most important sentences of a text using an algorithm and uses those original sentences to create the auto-summary (unlike generative AI). As the constituted sentences are machine selected, they may not fully reflect the body of the work, so we strongly advise that the original content is read and cited. The auto generated summaries were curated by the editor to meet Springer Nature publication standards. To cite this content, please refer to the original papers.

Machine generated keywords: propagation, clock, buffer, propagation delay, subthreshold, current, peak, interconnect, delay, insertion, interconnect delay, reduction, line, insert, tree

Performance Analysis for Reliable Nanoscaled FinFET Logic Circuits [1]

This is a machine-generated summary of:

Mushtaq, Umayia; Sharma, Vijay Kumar: Performance analysis for reliable nanoscaled FinFET logic circuits [1]

Published in: Analog Integrated Circuits and Signal Processing (2021)

Link to original: https://doi.org/10.1007/s10470-020-01765-z

If you want to cite the papers, please refer to the original.
For technical reasons we could not place the page where the original quote is coming from.

Abstract-Summary

"Fin-shaped field effect transistor (FinFET) is an important device which uses the concept of multi-gates and it is not only scalable but also dissipate lower power at lower technology nodes."

"This paper designs a low leakage input dependent (INDEP) approach for FinFET devices at 16 nm technology node."

"Numbers of logic gates are designed and simulated with the help of MOSFET and FinFET devices."

"Simulation results for different logic circuits show the large reduction in leakage power in case of FinFET logic gates as compared to MOSFET devices and more leakage saving in case of INDEP approach as compared to conventional as well as LECTOR technique."

Introduction

"The scaling of threshold voltage increases the leakage current which limits the functionality of circuits [2]."

"The methods which reduce leakage power at circuit level include transistor stacking, multiple threshold voltage (Vth), multiple power supply (Vdd), dynamic Vdd and dynamic Vth [3]."

"Various techniques are used to reduce leakage current both at process and circuit levels."

"At circuit level, leakage can be controlled by controlling drain, gate, source and body (substrate) voltages, besides by threshold voltage as well."

"The leakage current decreases as the number of off-state stacking transistors are increased."

"LECTOR and INDEP leakage reduction approaches are the CMOS based special configurations of the transistors and explained as below: Continuous device scaling increases the sub-threshold leakage current and to reduce the sub-threshold leakage current, we can use LECTOR technique."

"Suitable selection of inputs at gate terminals of the INDEP transistors decreases large leakage current."

Comparison of Various Performance Metrics of MOSFET and FinFET

"The comparison of various performance metrics like leakage power dissipation, propagation delay and PDP are carried out in case of CMOS inverter, CMOS 2-input NAND (NAND2), CMOS 2-input NOR (NOR2), FinFET inverter, FinFET NAND2 and FinFET NOR2 gates."

"The results are verified by the circuit simulations [4, 5] using Mentor Graphics ELDO simulator at room temperature with power supply of 0.9 V. The channel lengths of N and P-FinFET devices are equal and 16 nm while the channel widths for N and P-FinFET devices are 32 nm and 64 nm respectively."

"The simulation results show that leakage power dissipation in FinFET inverter, FinFET NAND2 and FinFET NOR2 gates are decreased by 96.48%, 99.75% and 99.76% respectively in comparison to the CMOS inverter, CMOS NAND2 and CMOS NOR2 gates."

"These results clearly infer that leakage is lower in case of FinFET circuits as compared to CMOS logic circuits."

FinFET Logic Gates Design

"There are two leakage control FinFETs between pull-up and pull-down networks of logic gates in LECTOR with FinFET logic."

"One of the leakage control FinFET is always near its cut-off voltage for any combination of inputs that reduces leakage current."

"Leakage power dissipation is decreased by 20.85%, 15.42% and 36.83% respectively for FinFET LECTOR inverter, FinFET LECTOR NAND2 and FinFET LECTOR NOR2 gates as compared to conventional FinFET logic gates."

"Low leakage INDEP technique is applied to FinFET logic gates and observed the huge amount of reduction in leakage current."

"INDEP technique reduces the leakage power dissipation and provides rail to rail output voltage swing if the inputs to the gate terminals are appropriate based on Boolean logic."

"Leakage power dissipation is reduced by 68.33%, 58.87% and 65.70% in FinFET INDEP inverter, FinFET INDEP NAND2 and FinFET INDEP NOR2 gates respectively in comparison to the conventional FinFET logic gates."

Simulation Results and Discussion

"All the experimental data and results are obtained at 16 nm MG BSIM4 PTM technology node using ELDO simulator at room temperature with Vdd = 0.9 V. Circuits are tested for leakage power dissipation, propagation delay and PDP."

"The intermediate nodes are responsible for improver response at output in FinFET with LECTOR technique while in INDEP FinFET logic circuit, there is no loss of rail to rail swing voltage at output node."

"When we are increasing the supply voltage from 0.5 to 0.8 V, the power dissipation shows the measurable increase for both logics."

"When the width of transistors is doubled then power dissipation is increasing for both logic levels."

"Leakage power dissipation, propagation delay and PDP are calculated for ring oscillator at 16 nm technology node for FinFET, FinFET with LECTOR and FinFET with INDEP approaches."

Conclusion

"INDEP and LECTOR leakage reduction approaches are utilized for designing the logic circuits using FinFET devices and compared with the conventional designs using Mentor Graphics ELDO simulator at 16 nm technology node."

"It has been observed that INDEP approach provides large saving of the leakage power as compared to LECTOR and conventional designs."

"The leakage power dissipation is saved by 94.91% in case of INDEP approach while it is 74.65% in case of LECTOR approach as compared to conventional design for ring oscillator."

"The PDP is improved by 97.58% in INDEP approach while it is 84.66% in LECTOR technique as compared to conventional design."

High-Performance Radiation Hardened NMOS Only Schmitt Trigger Based Latch Designs [6]

This is a machine-generated summary of:

Shah, Ambika Prasad; Gupta, Neha; Waltl, Michael: High-performance radiation hardened NMOS only Schmitt Trigger based latch designs [6]

Published in: Analog Integrated Circuits and Signal Processing (2021)

Link to original: https://doi.org/10.1007/s10470-021-01924-w

If you want to cite the papers, please refer to the original.

For technical reasons we could not place the page where the original quote is coming from.

Abstract-Summary

"A soft error tolerant latch built on a Schmitt trigger, which is entirely realized with NMOS transistors with an additional voltage booster, which we refer to as NST-VB, is proposed."

"To evaluate the circuits' radiation resilience, we identify the most sensitive nodes by analyzing the critical charges at the various latches' internal sensitive nodes."

"The critical charge to power delay area product ratio (QPAR) is calculated and evaluated for the proposed latch's effectiveness compared to other considered latches."

Introduction

"With the improvements of the integrated devices, the amount of charges stored on a single node in a circuit has been significantly reduced, so also low energy protons could generate sufficient amount of charge by direct ionization to cause SEU."

"The radiation hardening methods for latches can be roughly classified into two categories [7]. (i) The hardened circuits are considered independent of the size of transistors and the node capacitances."

"This latch tolerates the soft error for its internal nodes without increasing the node capacitance and the transistor geometry."

"In the SEM latch, two additional transistors are added to the feedback inverter to achieve an Schmitt trigger (ST) arrangement to improve the critical charge at the internal node N2."

"This circuit relies on introducing an ST inverter [8] in place of the inverter I1 of the unhardened latch to shield the internal nodes against TF."

Soft Error Modeling

"It is a widely used parameter to analyze the impact of soft error issues on devices and circuits."

"A soft error can be generated by a current pulse with a high amplitude in the simulations with a short pulse duration."

"We evaluate the minimum duration and magnitude of the caused current pulse, which is needed to flip the output logic of a latch circuit."

"The soft error robustness of the circuits is analyzed with the soft error rate."

"At deep submicron technology, the circuits are highly sensitive to soft errors because smaller node capacitances and small critical charges are attributed to the lower supply voltages."

"The critical charge is used to analyze the soft error tolerance of the circuits."

Analysis of the Radiation Resilience of Inverters

"For our radiation hardening analysis of the inverters, we consider the circuits of a CMOS inverter, an ST inverter, and an NST-VB based inverter."

"As only NMOS transistors drive the NST-VB circuit's output, the only critical condition that needs to be considered is the input state logic '0', and the corresponding output state logic '1'."

"It has to be noted that the critical charge for the NST-VB circuit is 78.09% and 47.8% higher as compared to the CMOS and ST circuits, respectively."

"Results show that the critical charge increases with supply voltage for the investigated inverter circuits because the higher supply voltage provides more conduction current, which increases the charge at the sensitive node."

"It is also observed that the critical charge for the NST-VB inverter is higher than for the other circuits; hence it can be beneficial for designing latches with better radiation hardening than ST-based latch."

Proposed NST-VB Based Latch

"This improvement can be beneficial for the design of radiation-hardened latches and SRAM cells."

"We are utilizing the benefits of NST-VB against the radiation hardening to design the soft error tolerant latch, which can be of interest for applications in harsh environments."

"To improve the radiation hardening of the NST-VB circuit, an inverter is added to the transparent path of the latch."

"The NST-VB circuit provides better robustness to the soft errors due to the charges stored at the internal nodes x and y. For further radiation-hardened improvement, it would be conceivable that all three inverters in the unhardened latch can be replaced with the NST-VB inverter."

Evaluation and Comparison of Latches

"From results, it is observed that the critical charge for ST based latch [9] is higher than the SEM [10], M-SEM [11], and SIN-LC [12]; hence we considered unhardened latch, ST based latch, and proposed NST-VB based latch for further analysis."

"Further, the NST-VB latch has a higher critical charge than the unhardened latch and ST-based latch, indicating that the proposed latch has better soft error robustness."

"The result shows that the critical charge at the sensitive node of NST-VB latch is higher for all the process corners as compared to the unhardened latch and ST based latch."

"The result indicates that the scattering of node voltage is less for the proposed NST-VB latch as compare to the unhardened and ST based latches."

"Results demonstrate that the effect of process variation on the critical charge of NST-VB latch is less as compared to the ST based latch."

Conclusion

"We propose a new NST-VB based radiation-hardened latch."

"We clearly observe that the NST-VB inverter has better performance in terms of radiation hardening as compared to CMOS and ST inverters."

"We utilize the benefit of NST-VB inverter to implement the radiation-hardened latch."

"We analyze the critical charge, power dissipation, and delay of all the latches under the PVT variations, and NST-VB latch performs significantly better as compared to other latches."

"For the process variations analysis of different latches, we performed 5000 Monte Carlo simulations and observed that the failure probability of NST-VB latch is less as compared to the reference and ST latches."

Logically Optimal Novel 4:2 Compressor Architectures for High-Performance Applications [13]

This is a machine-generated summary of:

Priyadharshni, M.; Chathalingathu, Anishchandran; Kumaravel, S.; Manoharan, Arun; Veeramachaneni, Sreehari; Mahammad, Sk Noor: Logically Optimal Novel 4:2 Compressor Architectures for High-Performance Applications [13]

Published in: Arabian Journal for Science and Engineering (2020)

Link to original: https://doi.org/10.1007/s13369-020-04503-9

Copyright of the summarized publication:

If you want to cite the papers, please refer to the original.

For technical reasons we could not place the page where the original quote is coming from.

Abstract-Summary

"While the efficiency of multipliers relies on optimal reduction of partial products within constraint power-delay values, this can be achieved by 'compressors' a precisely designed special arithmetic module."

"The presented designs perform better than the contemporary designs in terms of latency, improved power-delay and area-delay product values."

"From the synthesis results, it can be seen that the proposed model provides 12.5–29.17% reduction in area and 7–14.23% reduction in propagation delay with respect to state-of-the-art 4:2 compressor designs."

"To estimate the performance stability in realistic applications, 4-bit, 8-bit and 16-bit Dadda multiplier integrated with proposed compressor cells is implemented and verified."

"The obtained synthesis results substantiate that Dadda multiplier integrated with proposed compressor cells performs better than conventional and state-of-the-art 4:2 compressor-based variants with 4–8% reduction in propagation delay, upto 23.370% reduction in area-delay product and upto 25.62% reduction in power-delay product for 4-bit multiplication, while 6.77–16.85% reduction in propagation delay, 2.62–18.70% reduction in area-delay product and 15.30–23.20% reduction in power-delay product are achieved for 8-bit multiplication."

"For 16-bit multiplication it shows upto 47.05% reduction in propagation delay, upto 21.68% reduction in area-delay product and upto 24.42% reduction in power-delay product for 16-bit multiplication."

Introduction

"The power- and delay-efficient multipliers are the key element in determining the performance of such architectures [14]."

"The optimal performance of multipliers relies on the reduction of partial products generated during multiplication process."

"Wallace had presented an efficient and faster reduction of partial products through carry save tree structure made from single-bit full adders."

"The performance efficiency of compressors over the conventional techniques gave it a wide acceptance in multiplier architecture [15–18]."

"The showcased model modifies the 4:2 compressor logical design for high-performance and energy-efficient 4:2 compressors."

"The results claim optimal power–delay product and area–delay product values, retaining architecture parallelism compared to its high-performance variant."

Compressor Architecture

"Multiplication process can be comprehended in three steps: (1) Generation of partial products. (2) Reducing partial products to carry save format. (3) Carry propagate addition of final two vectors obtained from step (2) above."

"Of the above three steps, being a temporal process, reduction of partial products demands significant computing resources."

"Delay optimal multiplier architecture development directly depends on the efficiency of partial production reduction block."

"Due to horizontal and vertical signal paths, compressors are simpler and benefitted by better regularity in tree structure than other compression technique [19, 20]."

4:2 Compressor Architecture

"The input–output relation of 4:2 compressor can be expressed mathematically as A 4:2 compressor is basically realized through combinations of multiplexers and exclusive OR logic [XOR]."

"Many researches are still going on at circuit level to achieve Pareto optimal 4:2 compressor design [21]."

"As the presented work points on performance acceleration of 4:2 compressor via logical decomposition, the analysis and comparisons are restricted to the works cited to gate-level optimization."

"The proposed model uses significant gate-level optimized architectures available from literature as benchmark for its performance evaluation, provided all the gates are realized in standard 65 nm CMOS logic cells by the synthesis tool."

Previous Works

"The author's prime aim to use 4:2 compressor was to benefit their multiplier tree with regular and symmetric structure for efficient routing and greater delay reduction with respect to usual CSA block-based tree."

"Not much focus had given for logical simplification to the formal 4:2 compressor design which resulted a bit longer critical path of four XOR gate delay."

"They were able to achieve the same critical path delay of 3 gate stage using pass transistor MUXs as basic building block that aids in power reduction."

"The authors replaced all gates of the 4:2 compressor circuitry by full-swing pass transistor logic-based MUX to achieve power efficient speed optimized compressor."

"Authors claim that their model gives 18% delay advantage in respect of full adder-based compressors [22]."

"HSPICE simulation justifies the choice by providing 5% reduction in latency for full adder-based model than the conventional design with a delay of 2.64 XOR gates on the critical path."

Proposed Work

"Based on referenced literature, it is evident that the critical path delay of an isolated 4:2 compressor is determined by the XOR gates of 'Sum' generation logic [23]."

"The proposed designs are able to achieve considerable reduction in ADP and PDP in contrast to contemporary efficient designs through carefully crafted logical equation blended with observations made from conventional 4:2 compressor truth table."

"The authors were keen in implementing compressor with least use of XOR logic, as XOR logic is susceptible to more delay and area."

"When compared to Hussin's compressor [24], the fastest gate-level compressor architecture is known present, the proposed compressor design costs significantly less power and silicon area utilization due to modest carry generation to the former."

"While counting on recent design variant by Arasteh and others, author's version excels in both area and delay thanks to modest sum–carry logic."

Emperical Results

"In case of standard compressor-based multiplier, authors designs perform significantly well with an advantage of 8% in delay, 20.22% and 23.37% reduction in PDP and ADP."

"The proposed 4-bit multiplier performs better than Hussin's fastest compressor-based multiplier in all aspects, i.e., delay, ADP and PDP values."

"The presented multiplier dominates multiplier build on standard benchmark 4:2 compressor by 23.52% improvement in delay, 20.22% depreciation in ADP, 2.07% depreciation in PDP values that uses proposed design 1 and 17.68% improvement in delay, 15.12% depreciation in ADP, 1.12% depreciation in PDP values that uses proposed design 2, respectively."

"With speedup of 35.29%, depreciation of 12.53% in ADP, 4.94% reduction in PDP values by proposed design 1 and a 29.41% of speedup, 7.42% decrease in ADP, 4.05% depreciation in PDP values by proposed design 2, the proposed 16-bit multipliers outshine Hussin's fastest 4:2 compressor-based multiplier."

Conclusion

"Authors examined various logically optimized 4:2 compressors from conventional design to state-of-the-art designs and proposed two new logical models that excel previous models in individual comparison and in applications."

"The proposed model is able to achieve considerable reduction in propagation delay thanks to the effective utilization of MUX logic over XOR in critical path."

"In 16-bit multiplication, multiplier with proposed designs achieves upto 47.05% reduction in propagation delay, upto 21.68% reduction in area–delay product and upto 24.42% reduction in power–delay product."

"The comparative analysis of implementation results motivates authors to conclude that the proposed 4:2 compressors shall be a desirable choice for the implementation of high-performance multipliers and other related applications."

Implementation of Low Power Null Conventional Logic Function for Configuration Logic Block [25]

This is a machine-generated summary of:

Rajasekar, P.; Subash Kumar, C. S.: Implementation of Low Power Null Conventional Logic Function for Configuration Logic Block [25]

Published in: Wireless Personal Communications (2019)

Link to original: https://doi.org/10.1007/s11277-019-06380-4

If you want to cite the papers, please refer to the original.

For technical reasons we could not place the page where the original quote is coming from.

Abstract-Summary

"Todays, transistor level design plays major impact in the power consumption of the VLSI based design."

"The various logic design models are used to reduce the power consumption that includes synchronous and asynchronous design model."

"Asynchronous circuits are used widely because it causes low noise, require less power."

"These asynchronous circuits are being implemented by Null Conventional Logic (NCL) which is a delay insensitive logic model."

"The Configuration Logic Block is the power consuming model in Field Programmable Gate Array (FPGA)."

"To reduce the power consumption of this module, we use the Differential Cascade Voltage Switch Logic (DVSL)."

Introduction

"This logic is an improved design model of Wave Dynamic Differential Logic (WDDL), Masked Logic using Fourier Transform (MLFT), two spacer alternating Dual rail circuit, Dual rail Random Switch Logic and Masked Dual rail pre-charge Logic."

"The IP blocks based system designs are plug and play assembly principles that leads to reduce the design time, testing etc So, the system architecture point of view, the asynchronous design is much easier design model rather than synchronous design to build the SOC."

"The NCL based asynchronous logic design cannot be neglected because of controlling the noise resistant, power efficient and improve the performance [26]."

"NCL is an asynchronous logic design that uses the interaction of alternating DATA and Null wave fronts, also known as delay insensitive paradigm."

"The synchronization is achieved by delay insensitive in NCL paradigm design and by handshaking signal in self-timed paradigm [27]."

"NCL design logic follows the monotonic transitions between DATA and NULL."

Related Works

"Current Mode logic (CML) and Dual-Rail Pre-charge logic (DRP) are existing logic used in synchronous circuits."

"Most of the semi-custom design flow model uses the dual rail pre-charge logic, because the power consumption is insensitive to the unbalanced load condition."

"Glitches are totally avoided an implementation of Masked Dual-Rail pre charge logic (MDPL) method, but in random masking method at the gate level design, the consumed power is randomized."

"Wave Dynamic Differential Logic [WDDL] [28] method requires only standard cells, so it reduces design time cost compared to full custom design where it fails to compensate the power imbalance."

"Dynamic Current Mode Logic (DyCML) [29] has been one of the best dynamic logic style models."

"To solve the unbalanced load capacitance problem the Three-Phase Dual-Rail pre charge logic (TDPL) [30] based on DRP style logic has been proposed."

NCL Design Model

"NCL logic, dual rail and quad rail signals eliminates the time reference logic."

"Before the output is de asserted all asserted input must be de asserted because of NCL logic gates are designed with hysteresis logic."

"NCL uses the dual rail which has two wires for single logic."

"NCL design logic is implemented in two different implementation, namely NCL-D and NCL-X. The former integrates completion detection into the dual rail while later relies separate detection completion circuit."

"The NCL dual rail logic function is expressed in terms Sum Of Product (SOP), for an example the AND gate function is defined as F = AB in Boolean logic where as in NCL Dual Rail $F = A'B + AB' + A'B'$."

"Function and complement functions are implemented using differential logics that use two output rails."

Simulation Results and Discussion

"The proposed NCL logic style is analyzed in basic TH_{23} NCL gate and derived 2–1 multiplexer."

"Further the proposed NCL logic style's power comparison has been analyzed in AND TH_{23} gate."

"The sample design has been implemented using different NCL logic style, and optimization, effectiveness analyzed in terms of number of transistor used."

"Various basic gates are implemented with different logic style."

"The proposed logic design is achieved as 70.58%, 60.57%, 68.65% and 2.8% of AND, OR, NAND, and NOR gates respectively in SDML design."

"SCMOS logic design consumes more power of 58.79%, 47.56%, 56.47% of AND, OR, and NAND gates against the proposed NCL design."

Conclusion

"The proposed DCLVL based NCL based logic gates can be implemented by less number of transistors compared to all other transistor model."

"In basic logic design consumes 15.68% power in AND gate as the best case and 41.81% power consumed in worst case."

"These basic logic gates model and multiplexer design modules are the fundamental building blocks of LUT and CLB in FPGA design."

Buffer Insertion for Delay Minimization in RLC Interconnects Using Cuckoo Optimization Algorithm [31]

This is a machine-generated summary of:

Karimi, Gholamreza; Moradi, Yavar: Buffer insertion for delay minimization in RLC interconnects using cuckoo optimization algorithm [31]

Published in: Analog Integrated Circuits and Signal Processing (2018)

Link to original: https://doi.org/10.1007/s10470-018-1318-y

If you want to cite the papers, please refer to the original.

For technical reasons we could not place the page where the original quote is coming from.

Abstract-Summary

"Cuckoo optimization algorithm (COA) as an efficient evolutionary algorithm is used for buffer insertion in digital circuits."

"Inductance effects on optimal number and size of buffer are investigated."

"Buffer insertion in RLC lines is done in case study-I. Buffer insertion in tree structured inductive interconnects is dealt in case study-II."

"Optimal area (size and number of buffers) is another result of paper that can lead to economical number of chips."

Introduction

"Buffer insertion has been shown an increasingly common technique to reduce long interconnect delay [32–37]."

"Number and size of buffers must be changed in order to get optimal solution, minimizing the total propagation delay from source to sink."

"Not only inductance effect should not be neglected but also the interconnect delay should be considered in VLSI circuits modeling and optimizing."

"Several researchers have been performed for buffer insertion within line [32–34, 36–39] and tree structured [35] interconnects, so that a main part of them have considered inductance role [2, 5, 10] to determine optimum number and size of buffers."

"In most previous works, classical and mathematical algorithms are introduced for buffer insertion and optimization problem."

"Evolutionary algorithms has played an important role in increasing automated design and optimization of integrated circuits."

"Optimization parameters such as power dissipation, delay, chip area can be considered as a function in VLSI design."

Cuckoo Optimization Algorithm

"Cuckoo optimization algorithm was introduced by Rajabioun [40]."

"The cuckoos do not build their own nests and choose the nest of another species to lay their eggs."

"Rest of the cuckoo eggs have the opportunity to grow up [41]."

"When the cuckoo eggs hatch earlier than the host eggs, the chick cuckoos throw out the host eggs."

"Young cuckoos grow up and mature and as soon as egg laying approaches, they migrate to better habitats."

"New habitats that are chosen have more food resources and also more similarity of host eggs to cuckoo eggs."

Case Study-I: Buffer Insertion in RLC Lines

"One popular form of the inverter gate is CMOS inverter."

"CMOS inverter acts as a buffer for the interconnect load [42]."

"To investigate inductance effects on the improvement in propagation delay offered by buffer insertion, various values of inductance per unit length are considered 0, 3, 6 and 9 (nH/mm)."

"In buffer insertion using the COA the propagation delay reduction is up to 59.78% compared with the unbuffered lines."

"With increasing in inductance per unit length, less improvement in propagation delay is achieved using buffer insertion."

Case Study-II: Buffer Insertion in RLC Trees

"In the proposed buffer insertion technique, each branch of a tree is uniformly buffered."

"The maximum path propagation delay from source to sink of a buffered tree is considered as cost function."

"COA is responsible for finding optimum values for the number and size of buffers for each branch in order to minimize the propagation delay."

"The Percentage of propagation delay reduction using buffer insertion method optimized by the COA as compared with the unbuffered cases as a main result represented in these tables."

"For both balanced and unbalanced trees, buffer insertion using the COA reduces the propagation delay in all of the case studies."

"As inductance of the interconnect wire is increased, less improvement in delay is achieved by buffer insertion technique."

Conclusions and Future Work

"The cuckoo optimization algorithm was applied to optimize buffer insertion in RLC interconnects in order to reduce the propagation delay."

"For unbuffered line, increasing inductance has a significant effect on delay reduction and buffer insertion can not compensate negative effect of inductance on delay well."

"For unbuffered trees, negative effect of increasing inductance on delay is low and buffer insertion in trees is an efficient method that causes significant reduction in delay."

"Inductance has a positive effect on buffer area reduction in all of the case studies."

"Buffer insertion failures lead to use the wire shaping as a more effective technique in advanced technologies with more inductance effect [43]."

Delay Analysis of Buffer Inserted Sub-threshold Interconnects [44]

This is a machine-generated summary of:

Dhiman, Rohit; Chandel, Rajeevan: Delay analysis of buffer inserted sub-threshold interconnects [44]

Published in: Analog Integrated Circuits and Signal Processing (2016)

Link to original: https://doi.org/10.1007/s10470-016-0860-8

If you want to cite the papers, please refer to the original.

For technical reasons we could not place the page where the original quote is coming from.

Abstract-Summary

"An analysis of interconnect delay minimization by CMOS buffer insertion in sub-threshold regime is presented."

"Analytical expressions are developed to calculate the total delay and optimum number of buffers required for delay minimization in sub-threshold interconnects."

"Considering delay minimization by buffer insertion, the effects of voltage-scaling on the delay and optimum number of buffers have been analyzed."

"It is demonstrated that voltage scaling in sub-threshold regime reduces the number of buffers required to attain the minimum delay."

Introduction

"Improvements in the technology have enlarged circuit die sizes and the number of transistors, thereby increasing the length and hence the delay of on-chip interconnects."

"Driving CMOS interconnect circuits in sub-threshold logic provides orders of magnitude power reduction over conventional super-threshold logic, but causes an exponential increase in the propagation delay."

"Interconnect delay thus becomes more problematic in sub-threshold VLSI systems."

"Propagation delay of an on-chip interconnect can be reduced by inserting a number of buffers along its length [45, 46]."

"Minimization of interconnect delay in sub-threshold logic by buffer insertion and its analysis is still the topic of research [47, 48]."

"The present work deals with analytical studies for the minimization of sub-threshold interconnect delay by buffer insertion."

"A method is also presented to determine the optimum number of buffers required for the minimum delay based on the analytical expressions derived using sub-threshold current model."

Sub-threshold Timing Analysis

"In VLSI, it is a common practice to drive interconnects by CMOS buffers."

"CMOS inverter acts as a buffer for the interconnect load [49]."

"Inv1 and Inv2 are the driving and driven gates respectively, where R is the interconnect resistance and C is the interconnect capacitance and includes the input gate capacitance of Inv2."

"Interconnect is modeled as lumped resistive-capacitive in order to emphasize the exponential behaviour of the MOS device."

Determination of Delay in Buffer Inserted Interconnects

"When buffers are inserted, the interconnect capacitance (C/n) of each subsection comes in parallel with the capacitance (C_{buf}) of the following buffer stage."

"The falling output waveform of the first stage is input to the following buffer stage."

"The term $t_{first\ stage}$ of t_{90} in (13) describes the time required for the output voltage of first buffer to switch from V_{DD} to V_n assuming that output voltage is initially at V_{DD}."

"The term $t_{int.\ stage}$ is defined as the time required for intermediate buffer stages between the first and last stage to transit from V_{DD}–V_p to V_n or vice versa."

"These two expressions are alternately added to the total delay for each corresponding buffer stage up to the input of the final stage of the buffer chain."

"The delay expression for the falling output of an intermediate buffer is expressed asThe delay of rising buffer output, i.e., t_p is of the similar form."

Results and Discussion

"N_o is 8 when supply voltage is 650 mV, while n_o is 4 at a scaled voltage of 350 mV. The basic principle of interconnect delay minimization by buffer insertion is based on the equalization of buffer delay and interconnect delay."

"The decrease of current drive with voltage scaling increases delay."

"At scaled voltages, although optimal n is small but delay is relatively higher."

"The percentage improvement in energy dissipation by scaling the supply voltage from 650 to 350 mV is about 45%."

"For either case, the sub-threshold voltage scaling is limited to the requirement as demonstrated in [50] for proper operation of the CMOS buffers."

"In the present context, it is of interest to compare the effect of voltage scaling on the PDP of buffer loaded sub-threshold RC interconnects."

"V_{DD} value of 330 mV is obtained as the optimum voltage for minimum power-delay product."

Process Variabilty in Sub-threshold Interconnects

"This is because sub-threshold circuits are highly sensitive to device threshold voltage which exponentially affects the on-current."

"Variation in the threshold voltage causes different relative strength of n-channel and p-channel devices and thus affects functionality, delay and power dissipation of logic gates [51]."

"It is important to understand and investigate the influence of process variations under sub-threshold regime for the prediction of circuit performance."

"It is noted that the sensitivity of delay against threshold voltage variations is 410.65%."

"It can be observed that power dissipation in the sub-threshold regime is most sensitive to the variation in effective gate length since reduction in L_n decreases the threshold voltage and sub-threshold current rises exponentially."

"The sensitivity of power against effective gate length variations is 513%."

Conclusions

"Considering delay minimization by buffer insertion, the effects of voltage scaling on delay and optimum number of buffers have been analyzed."

"Analysis shows that optimum number of buffer and power dissipation decrease with voltage scaling."

"The application of delay expression developed in this paper demonstrates that optimum number of buffer reduces by 42% under the effect of voltage-scaling, however at the cost of increased delay."

"Monte Carlo analysis also shows that there is a significant increase in delay and power dissipation when $\pm 3\sigma$ variations in the threshold voltage and channel length, respectively occur."

Clock Skew Optimization for Peak Current Reduction [52]

This is a machine-generated summary of:

Benini, L.; Vuillod, P.; Bogliolo, A.; Micheli, G. De: Clock Skew Optimization for Peak Current Reduction [52]

Published in: Journal of Signal Processing Systems (1997)

Link to original: https://doi.org/10.1023/a:1007982806152

If you want to cite the papers, please refer to the original.

For technical reasons we could not place the page where the original quote is coming from.

Abstract-Summary

"The presence of large current peaks on the power and ground lines is a serious concern for designers of synchronous digital circuits."

"We propose a methodology for reducing the amplitude of the current peaks."

"We propose an algorithm that, for a given clock cycle time, determines the clock arrival time at each flip-flop in order to minimize the current peaks while respecting timing constraint."

"Our results on benchmark circuits show that current peaks can be reduced without penalty on cycle time and average power dissipation."

References

1. Mushtaq U, Sharma VK (2021) Performance analysis for reliable nanoscaled FinFET logic circuits. Analog Integr Circ Sig Process. https://doi.org/10.1007/s10470-020-01765-z
2. Zimpeck AL, Meinhardt C, Reis RA (2015) Impact of PVT variability on 20 nm FinFET standard cells. Microelectron Reliabil 55(9–10):1379–1383
3. Roy K, Mukhopadhyay S, Mahmoodi-Meimand H (2003) Leakage current mechanisms and leakage reduction techniques in deep-submicrometer CMOS circuits. Proc IEEE 91(2):305–327
4. Sun K, Mou S, Qiu J, Wang T, Gao H (2018) Adaptive fuzzy control for non-triangular structural stochastic switched nonlinear systems with full state constraints. IEEE Trans Fuzzy Syst 27(8):1587–1601
5. Qiu J, Sun K, Wang T, Gao H (2019) Observer-based fuzzy adaptive event-triggered control for pure-feedback nonlinear systems with prescribed performance. IEEE Trans Fuzzy Syst 27(11):2152–2162
6. Shah AP, Gupta N, Waltl M (2021) High-performance radiation hardened NMOS only Schmitt Trigger based latch designs. Analog Integr Circ Sig Process. https://doi.org/10.1007/s10470-021-01924-w
7. Nan H, Choi K (2011) Novel radiation hardened latch design considering process, voltage and temperature variations for nanoscale CMOS technology. Microelectron Reliabil 51(12):2086–2092
8. Melek LAP, da Silva AL, Schneider MC, Galup-Montoro C (2017) Analysis and design of the classical CMOS Schmitt trigger in subthreshold operation. IEEE Trans Circ Syst I Regul Pap 64(4):869–878
9. Lin S, Kim Y-B, Lombardi F (2010) Design and performance evaluation of radiation hardened latches for nanoscale CMOS. IEEE Trans Very Large Scale Integr Syst 19(7):1315–1319
10. Sasaki Y, Namba K, Ito H (2006) Soft error masking circuit and latch using Schmitt trigger circuit. In: 1st IEEE international symposium on defect and fault tolerance in VLSI systems. IEEE, pp 327–335
11. Lin S, Kim Y-B, Lombardi F (2009) Soft-error hardening designs of nanoscale CMOS latches. In: 27th IEEE VLSI test symposium. IEEE, pp 41–46
12. Omaña M, Rossi D, Metra C (2007) Latch susceptibility to transient faults and new hardening approach. IEEE Trans Comput 56(9):1255–1268

13. Priyadharshni M, Chathalingathu A, Kumaravel S, Manoharan A, Veeramachaneni S, Mahammad SKN (2020) Logically optimal novel 4:2 compressor architectures for high-performance applications. Arabian J Sci Eng. https://doi.org/10.1007/s13369-020-04503-9
14. Balobas D, Konofaos N (2018) Low-power high-performance CMOS 5-2 compressor with 58 transistors. Electron Lett 54(5):278
15. Kwon O, Nowka K, Swartzlander EE (2002) A 16-bit by 16-bit MAC design using fast 5:3 compressor cells. J VLSI Signal Process Syst Signal Image Video Technol 31(2):77
16. Oklobdzija VG (2000) High-speed VLSI arithmetic units: adders and multipliers. In: Chandrakasan A, Bowhill WJ, Fox F (eds) Design of high-performance microprocessor circuits. IEEE Press, pp 181–204
17. Santoro MR, Horowitz MA (1989) SPIM: a pipelined 64*64-bit iterative multiplier. IEEE J Solid-State Circuits 24(2):487
18. Nagamatsu M, Tanaka S, Mori J, Hirano K, Noguchi T, Hatanaka K (1990) A 15-ns 32*32-b CMOS multiplier with an improved parallel structure. IEEE J Solid State Circuits 25(2):494. https://doi.org/10.1109/4.52175
19. Oklobdzija VG, Villeger D, Liu SS (1996) A method for speed optimized partial product reduction and generation of fast parallel multipliers using an algorithmic approach. IEEE Trans Comput 45(3):294. https://doi.org/10.1109/12.485568
20. Bickerstaff KC (2007) Optimization of column compression multipliers. Docotoral Dissertation, Department of Electrical and Computer Engineering, University of Texas, Austin
21. Arasteh A, Moaiyeri MH, Taheri M, Navi K, Bagherzadeh N (2018) An energy and area efficient 4:2 compressor based on FinFETs. VLSI J Integr 60:224
22. Ohkubo N, Suzuki M, Shinbo T, Yamanaka T, Shimizu A, Sasaki K, Nakagome Y (1995) A 4.4 ns CMOS 54/spl times/54-b multiplier using pass-transistor multiplexer. IEEE J Solid State Circuits 30(3):251
23. Lin HL, Chang RC, Chan MT (2004) Design of a novel radix-4 booth multiplier. In: The 2004 IEEE Asia-Pacific Conference on Circuits and Systems, 2004. Proceedings, vol 2. IEEE, pp 837–840
24. Hussin R, Shakaff AYBM, Idris N, Sauli Z, Ismail RC, Kamarudin AA (2008) An efficient modified booth multiplier architecture. In: 2008 International Conference on Electronic Design. IEEE, pp 1–4
25. Rajasekar P, Subash Kumar CS (2019) Implementation of low power null conventional logic function for configuration logic block. Wireless Personal Commun. https://doi.org/10.1007/s11277-019-06380-4
26. Bucci M, Giancane L, Luzzi R, Trifiletti A (2006) Three-phase dual-rail pre-charge logic. In: Cryptographic hardware and embedded system CHES 2006, vol 4249. International Association for Cryptologic Research, Yokohama, pp 232–241
27. Al-Assadi W, Kakarla S (2009) Design for test of asynchronous NULL convention logic (NCL) circuits. J Electron Testing 25(1):117–126
28. Lee H, Kim Y (2013) Low power null convention logic circuit design based on DCVSL. In: IEEE 56th international midwest symposium on circuits and systems (MWSCAS). Ohio Union, IEEE, pp 29–32
29. Ma J, Kapoor HK, Krilavicius T, Man KL, Zhang N, Lim EG, Jeong TT, Guan SU, Seon JK (2011) Specification and analysis of NCL circuits. Eng Lett 19(3):215–222
30. Datta D, Mitra P, Sen A (2013) Low power configuration logic block design using asynchronous static. Int J Soft Comput Eng 3(1):260–263
31. Karimi G, Moradi Y (2018) Buffer insertion for delay minimization in RLC interconnects using cuckoo optimization algorithm. Analog Integr Circ Sig Process. https://doi.org/10.1007/s10470-018-1318-y
32. Ismail YI, Friedman EG (2000) Effects of inductance on the propagation delay and repeater insertion in VLSI circuits. IEEE Trans Very Large Scale Integr Syst 8(2):195–206
33. Alpert CJ, Devgan A, Quay ST (1999) Buffer insertion for noise and delay optimization. IEEE Trans Comput Aided Des Integr Circ Syst 18(11):1633–1645

34. Chu C, Wong DF (1997) A new approach to simultaneous buffer insertion and wire sizing. In: Proceedings IEEE/ACM international conference on computer-aided design. IEEE, pp 614–621
35. Ismail YI, Friedman EG (2001) Repeater insertion in tree structured inductive interconnect. IEEE Trans Circuits Syst 48(5):471–481
36. Chu C, Wong DF (2001) Closed form solution to simultaneous buffer insertion/sizing and wire sizing. ACM Trans Des Automat Electron Syst 6:343–371
37. Karimi G, Ahmadi A (2014) Buffer insertion for delay minimization using an improved PSO algorithm. Appl Math Inform Sci 8(5):2277–2285
38. Van Ginneken LPPP (1990) Buffer placement in distributed RC-tree networks for minimal Elmore delay. In: Proceedings of the IEEE international symposium on circuits and systems, vol 2. IEEE, pp 865–868
39. Ismail YI, Friedman EG (1998) Optimum repeater insertion based on a CMOS delay model for on-chip RLC interconnect. In: Proceedings of the eleventh annual IEEE international ASIC conference. IEEE, pp 369–373
40. Rajabioun R (2011) Cuckoo optimization algorithm. Appl Soft Comput 11(8):5508–5518
41. Mellal MA, Williams EJ (2015) Cuckoo optimization algorithm for unit production cost in multi-pass turning operations. Int J Adv Manuf Technol 76(1–4):647–656
42. Dhiman R, Chandel R (2017) Delay analysis of buffer inserted sub-threshold interconnects. Analog Integr Circ Sig Process 90(2):345–445
43. El-Mousry MA, Friedman EG (2007) Wire shaping of RLC interconnects. Integr VlSI J 40(4):461–472
44. Dhiman R, Chandel R (2016) Delay analysis of buffer inserted sub-threshold interconnects. Analog Integr Circ Sig Process. https://doi.org/10.1007/s10470-016-0860-8
45. Deodhar VV, Davis JA (2008) Optimal voltage scaling, repeater insertion, and wire sizing for wave-pipelined global interconnects. IEEE Trans Circ Syst I Regul Pap 55(4):1023–1030
46. Bakoglu HB, Meindl JD (1985) Optimal interconnection circuits for VLSI. IEEE Trans Electron Dev 32(5):903–909
47. Kil J, Gu J, Kim CH (2008) A high-speed variation-tolerant interconnect technique for sub-threshold circuits. IEEE Trans Very Large Scale Integr Syst 16(4):456–465
48. Dhiman R, Chandel R (2015) Compact models and computation of crosstalk for sub-threshold interconnect circuits. Analog Integr Circ Sig Process 82(3):637–652
49. Agrawal Y, Chandel R (2016) Crosstalk analysis of current-mode signaling-coupled RLC interconnects using FDTD technique. IETE Tech Rev 33(2):148–159
50. Calhoun BH, Chandrakasan AP (2005) Ultra-dynamic voltage scaling (UDVS) using sub-threshold operation and local voltage dithering. IEEE J Solid State Circuits 41(1):238–245
51. Kwong J, Chandrakasan A (2006) Variation-driven device sizing for minimum energy sub-threshold circuits. In: Proceedings of international symposium on low power electronics and design. IEEE
52. Benini L, Vuillod P, Bogliolo A, Micheli G (1997) Clock skew optimization for peak current reduction. J Signal Process Syst. https://doi.org/10.1023/a:1007982806152

Chapter 7
VLSI Circuits and Approximate Computing

Introduction by the Editor

This chapter discusses VLSI circuits and approximate computing, which are the upcoming fields in VLSI design. This chapter focuses on Tunable Memristor emulators. This is a CNFET based tuneable memristor emulator. It also discusses a VDTA based robust electronic lead tunable memristor emulator circuit along with methods that focus on analysis and characterization of leakage reduction. A multi bit soft error immune SRAM cell for low power application is also discussed in this chapter using BitMAC. Along with these a bit serial computation based efficient multiply accumulate unit. For a deep neural network accelerator is discussed. Design of low power multiplier with less area using quaternary carry. Incremental adder for newfangled processor. After that, an ultra-low power and fast voltage level shifter using Mullar C-cell for VLSI system is also discussed.

One of the states of art technique discuss in this chapter is carbon nanotube field effect transistor (CNFET) based memristor emulator circuit. The experiments have variables like Temperature variation and process variation. This study provides the design and alternative to the carbon nanotube field effect transistor. Along with this, a variant of voltage differencing transconductance amplifier is also discussed. Moreover, memristor emulator circuit and its variable performance is also studied. Mathematical model for memristor emulator circuit. Providing a mathematical model Is the main aim for this chapter and its further use of describing the characteristics and its new features. Analysis done in the experiments mentioned in this chapter suggests that the memristor emulator circuit is robust in nature. Major things discussed in this sharp there is a deep analysis characterization of the CMOS process. Different inverter topologies coupled with various leakage reduction methods have also been elaborated well in this chapter. Power leakage optimization needs to be done at every abstract level of VLSI design. This includes architectural level, technology level, circuit level, gate level, etc. All the comparison shown in

A. S. Shastri (ed.), *Optimization Methods in VLSI Design*,
https://doi.org/10.1007/978-981-95-2457-0_7

this chapter. Regards to leakage reduction, active power consumption and the speed of the circuit. Chapter also contains a detailed discussion on a multi-bit SRAM cell which is soft error immune, and which is used for low power applications. The discussion highlights the removal of cell with half select disturbance associating with Involving a special gate to drive the access transistor. Highlights the employment of column interleaving structure and error correction coding. Reducing the multi bit problem and soft error immunity Enhancement.

A deep neural network accelerator with multiply accumulate unit is discussed in this chapter in detail. The objective of this work is to save leakage power using the power gated design of the proposed architecture. The experimentation mentioned in this chapter involves design architecture for power gated bit cell and bit cell-based design of multiply accumulate unit for deep neural network. Modified multiply accumulate block architecture is compared with other established architectures on basis of area, power and critical path delay. The advantage of this approach is the coarse grain power gating technique. The discussion moves along the relative impact of the increase in the 8-bit multiplier accumulate block. Nudes that the design discussed in this chapter is much smaller than Single power gated bit cell an exhibits Rewarding performance compared with the state-of-the-art architectures. The chapter further elaborates the simulation and performance analysis of Muller C-based level shifters. Moreover, this chapter also discusses about fast and ultra-low power Muller C-based level shifter which perform level shifting to a wide range of output and input voltage.

Machine Generated Summaries

Disclaimer: The summaries in this chapter were generated from Springer Nature publications using extractive AI auto-summarization: An extraction-based summarizer aims to identify the most important sentences of a text using an algorithm and uses those original sentences to create the auto-summary (unlike generative AI). As the constituted sentences are machine selected, they may not fully reflect the body of the work, so we strongly advise that the original content is read and cited. The auto generated summaries were curated by the editor to meet Springer Nature publication standards. To cite this content, please refer to the original papers.

Machine generated keywords: operation, voltage, leakage power, multiplier, static, leakage, resistance, active, resistor, static power, ultralow, ground, charge, analog, network

A Novel CNFET Based Tunable Memristor Emulator [1]

This is a machine-generated summary of:

Aishwarya, Nilay; Nayak, Abhijeet; Pal, Indrajit; Kumar, Vikash; Islam, Aminul: A novel CNFET based tunable memristor emulator [1]

Published in: Microsystem Technologies (2019)

Link to original: https://doi.org/10.1007/s00542-019-04486-0

If you want to cite the papers, please refer to the original.

For technical reasons we could not place the page where the original quote is coming from.

Abstract-Summary

"This paper presents a novel carbon nanotube field effect transistor (CNFET) based memristor emulator circuit."

"The proposed memristor emulator circuit utilizes voltage differencing transconductance amplifier (VDTA) as an active building block."

"The operation of the proposed emulator circuit, as governed by the established equations, has been verified by performing simulations using 32-nm Stanford CNFET model."

"Further, the robustness of the emulator circuit has been investigated by subjecting it to process and temperature variations."

Introduction

"The memristor's resistance depends on the voltage applied across its terminals or the amount of charge that has flowed through it."

"From [2] we can obtain a relationship between the memristor terminal voltage (v) and current (i), i.e., An ideal memristor is a passive two-terminal electronic device that expresses the property of memristance (just as a resistor expresses resistance and an inductor expresses inductance)."

"The characteristic hysteresis loop of these memristor emulators pinch-off at lower frequencies, thereby limiting the circuital operations."

"Further, voltage differencing transconductance amplifier (VDTA) is used as the active building block for realizing the proposed memristor emulator circuit."

"The proposed emulator circuit possesses characteristics such as electronic tunability, improved range of frequency of operation, lower power consumption, etc The proposed memristor emulator design utilizes two VDTA active elements, three grounded passive elements (i.e., two grounded resistors, a grounded capacitor) and an active analog voltage multiplier."

Carbon Nanotube Field Effect Transistor (CNFET)

"Ballistic transport of charge carriers makes CNFET one of the most promising devices [3] to be considered as an alternative to MOSFET."

"With ultra-long (~1 μm) mean-free-path of charge carriers while moving through intrinsic CNT, ballistic or near-ballistic transport occurs in CNFET resulting in optimum device performance."

"The carbon nanotube (CNT) in CNFET is surrounded by high dielectric constant (κ) gate oxide (HfO_2) with thickness of 4 nm and dielectric constant of 16."

"The structure is called multi-walled carbon nanotube (MWCNT) if multiple CNTs with varying diameter are rolled concentrically inside one another, whose diameter varies from several nm to tens of nm."

"The carbon nanotubes can either be metallic (conductor) or semiconducting depending on the angle (denoted by chirality integer vector (n_1, n_2) in which they are rolled."

Circuit Description of VDTA

"The differential-input differential-output transconductance amplifier of the first stage has two high impedance input terminals 'p' and 'n'; the terminal 'z' generates an output current (i_z) which is proportional to the difference in the voltages at two input terminals 'p' and 'n' (i.e., v_p–v_n)."

"The voltage drop at the 'z' (v_z) terminal produces output current at the output terminals 'x^+' and 'x^-' of the second transconductance stage."

"The mathematical equations that govern the relationship between the terminals are given as follows,where g_{mf} and g_{ms} are the first and second transconductances of the VDTA respectively and are cited as Arbel–Goldminz transconductances."

The Proposed Memristor Emulator Circuit

"The resistance of the proposed emulator circuit changes with the change in the applied voltage, v_{in}."

"The Gilbert cell architecture based multiplier circuit is composed of two N-CNFET active loads, three N-CNFET differential pairs, and one active biasing resistor."

"Transistors M_3, M_4 behave like a tail current source which mirrors the current drawn from supply voltage (V_{DD}) through transistors M_1, M_2, which are active biasing resistors."

"The voltage signals ('v_x' and 'v_y') to be multiplied are applied to the input terminals 'x_p', 'x_n' and 'y_p', 'y_n'."

Simulation Results and Discussion

"The circuit is biased with the supply voltage of $V_{DD} = V_{SS} = \pm 0.6$ V. While performing simulation analyses, all the VDTAs of the emulator have been biased with i_{BIAS} of 30 μA. The distinctive trait of a memristive circuit is that it yields a pinched hysteresis loop as its current–voltage characteristics, when subjected to an alternating voltage signal [4]."

"Hysteresis loops are formed because the current through the memristor does not vary linearly with the applied voltage; unlike a resistor that obeys Ohm's law A similar attribute is observed for the proposed memristor emulator and such observations are discussed below."

"The proposed emulator was subjected to a sinusoidal excitation of amplitude 30 μA and frequency 10 kHz in order to examine the operation of the circuit."

"In order to obtain frequency range of operation of the memristor emulator, the frequency 'f', was varied from 1 to 500 kHz."

Variability Performance of Proposed Memristor Emulator

"In order to assess the robustness of the proposed VDTA based memristor emulator design, a reliability analysis has been conducted by subjecting various parameters critical for circuital operations to variations."

"The reliability of the proposed emulator has been assessed by subjecting the design to process and environmental variations."

"Taking into consideration the interdependence of the aforementioned device parameters, the overall device parameter fluctuations is approximated by applying Gaussian distribution with 3σ variations of ±10% only on the device threshold voltage V_t [5]."

"Variations in the environment occur due to fluctuating environmental parameters."

"Such variations may translate into fluctuations in critical circuital parameters, thus hampering the operation of the circuit."

"Such fluctuations may impact device parameters like carrier mobility (μ_N) and threshold voltage (V_t), thus affecting circuital parameters and operations causing thermal instability."

Conclusion

"The featured memristor emulator proposes a simple circuit-topology, with fully-integrated structure, low power consumption, and tuneable capability by which the characteristics of the memristor can be varied electronically."

"The mathematical model for the proposed memristor emulator circuit has been established, and has been used to describe the characteristics and novel features of the circuit."

"The various analyses performed in this paper suggest that the proposed memristor emulator circuit is robust and variation-tolerant at deep-submicron technology nodes."

A VDTA-Based Robust Electronically Tunable Memristor Emulator Circuit [6]

This is a machine-generated summary of:

Pal, Indrajit; Kumar, Vikash; Aishwarya, Nilay; Nayak, Abhijeet; Islam, Aminul: A VDTA-based robust electronically tunable memristor emulator circuit [6]

Published in: Analog Integrated Circuits and Signal Processing (2019)

Link to original: https://doi.org/10.1007/s10470-019-01575-y

If you want to cite the papers, please refer to the original.

For technical reasons we could not place the page where the original quote is coming from.

Abstract-Summary

"A fully-integrated tunable grounded memristor emulator circuit based on voltage differencing transconductance amplifier (VDTA) has been proposed."

"The proposed memristor emulator circuit utilizes two VDTA active building blocks, two grounded resistors, a grounded capacitor and a four-quadrant analog multiplier."

"The operations of the proposed emulator circuit, as governed by the established model, have been verified by performing simulations in Cadence Virtuoso at 45 nm technology node."

Introduction

"The TiO_2-based memristor, as described by Strukov and others [7] is uniquely characterized by a frequency-dependent pinched hysteresis loop in its I–V plot for sinusoidal excitations."

"Significant research work has been observed in the area of memristor emulator circuits which can mimic the unique properties of the TiO_2 memristor and are also physically realizable [8–19]."

"Authors in these papers have proposed new memristor emulator circuits by using active building blocks (ABBs) and a number of passive elements."

"These emulators are lacking in certain novel features like, electronically tunable characteristics, occurrence of pinching of the hysteresis loop at relatively low-frequencies, range of operation and low power consumption, etc This paper presents a memristor emulator using CMOS-based voltage differencing transconductance amplifier (VDTA) as an active building block, which imparts characteristics like electronic tunability, improved range of frequency of operation, lower power consumption."

Circuit Description of VDTA

"The advantages of utilizing VDTA as an active device are as follows: (a) VDTA does not require the use of external resistors and are substituted by internal transconductances, (b) it exhibits two different values of transconductances, which are electronically controllable by external bias currents, (c) It has the capability of operation in voltage mode, current mode as well as transconductance mode as compared to other active devices reported (d) All the ports of VDTA exhibit high impedance, (e) VDTA-based filters allow independent controllability of resonant frequency and quality factor as compared to other active devices (f) VDTA-based designs also possess low active and passive component sensitivities."

"The voltage drop at the 'z' (v_z) terminal produces output current at the terminals 'x^+' and 'x^-' of the second transconductance stage (T-II)."

The Proposed Memristor Emulator Circuit

"The proposed emulator circuit emulates the change in the resistance of the memristor with the change in the applied input signal."

"The proposed design uses a MOS-based four quadrant analog multiplier circuit which is based on Gilbert cell architecture [20]."

"The multiplier circuit based on Gilbert cell architecture is composed of three NMOS differential pairs, two NMOS active loads and one active biasing resistor."

"Voltage signals ('v_x' and 'v_y') to be multiplied are applied to input terminals 'x_p', 'x_n' and 'y_p', 'y_n' respectively."

"For $VDTA_1$, the current at the 'x' terminal (i_{x1}) is given by,where v_{fb} is the feedback voltage from the analog multiplier."

Analysis and Results

"The proposed memristor emulator is simulated in Virtuoso Analog design Environment of Cadence at CMOS 45-nm CMOS process node."

"The G_M and the frequency range of operation of the proposed memristor emulator can be tuned by the bias current."

"This section presents the observations obtained from the experimental realization of the proposed memristor emulator circuit."

"The experimental results presented here, suggests the memrisitve characteristics of the proposed VDTA based emulator circuit with variation in operating frequency."

"Keeping in view the importance to mitigate the impact of device variability, a study has been conducted to analyze the robustness of the proposed VDTA-based memristor emulator design."

"A limited spread in memductance value of the emulator circuit translates into a narrow spread in other design metrics of the memristor emulator, thereby validating the proposed design to be variation aware."

Conclusion

"A novel memristor emulator circuit utilizing VDTA as active building elements is presented."

"The circuit presented offers a simple circuit-topology, with fully integrated configuration, low power consumption, and tunable capability by which the characteristics of the emulated memristor can be varied electronically."

"The mathematical model for the proposed memristor emulator has been established in order to explain the characteristics and novel features of the circuit."

Analysis and Characterization of Leakage Reduction Methodologies for Stacking, Body Biasing and DLS in 65 nm CMOS Technology [21]

This is a machine-generated summary of:

Kilani, Dima; Mohammad, Baker; Alhawari, Mohammad; Saleh, Hani; Ismail, Mohammed: Analysis and characterization of leakage reduction methodologies for stacking, body biasing and DLS in 65 nm CMOS technology [21]

Published in: Analog Integrated Circuits and Signal Processing (2019)

Link to original: https://doi.org/10.1007/s10470-019-01520-z

If you want to cite the papers, please refer to the original.

For technical reasons we could not place the page where the original quote is coming from.

Abstract-Summary

"This paper presents analysis, characterization and measurement results in a 65 nm CMOS process for different inverter topologies that use various leakage reduction methodologies."

"The measurement results are carried out to compare the topologies in regards of leakage reduction, active power and speed."

"The measurement results indicate that the DLS is an excellent option for ultra-low power IoTs where the leakage is minimized, however, it has a limitation in terms of performance."

"Stacking and body biasing support low leakage power and moderate performance."

Introduction

"Reducing leakage power is a key requirement in any power-constrained IoTs to achieve long lasting battery lifetime."

"Several solutions have been reported in the literature to reduce leakage power at different levels of abstraction such as architecture, technology, and circuit levels."

"Power gating degrades the performance during the active mode due to the IR drop across the sleep transistor."

"The sleep transistor size has an impact on the transition time between different power modes."

"Power gating reduces the leakage current but does not eliminate it because the sleep transistor has its own leakage."

"A super cutoff power gating is introduced where a reverse bias voltage is applied to the gate of the sleep transistor [22]."

"The measurement results indicate that the DLS technique is an effective way to reduce the leakage power for applications that are always on and operating at low frequency."

Leakage Reduction Methodologies for CMOS

"To better analyze the impact of various leakage reduction methodologies, a simple inverter gate has been selected."

"Even though the leakage reduction study focuses on the inverter gate, the study can be applied to any CMOS gate."

"The leakage current has been reduced by 5× when increasing the body voltage from 1.2 to 1.6 V. A Further increase in the body voltage will not have a high impact on leakage reduction."

"The reduction in leakage with body bias should be higher than the power consumed by the charge pump."

"Each configuration has been implemented as a series of 100 inverter cells with fan-out of 1 in order to have the ability to measure the leakage and active currents."

Measured Results

"The impact of the supply voltage scaling on the leakage saving is more significant for the DLS inverter rather than the other configurations."

"This is because stacking aims to reduce the leakage current where it is dominant at low frequency."

"As the clock frequency increases, the total current of the 2-stacked inverter and 2-stacked with body biasing configuration gradually increases and becomes equal to the total current of the standard inverter at f = 100 kHz."

"The output of the DLS configuration is controlled by the leakage current of the upper and lower transistors which becomes less dominant at high frequency."

"A standard topology is a good option for high performance devices operating at MHz frequency range but it has the highest leakage."

"Stacking and body biasing support low leakage power and intermediate workload where the device performance is needed but does not have to be high."

Conclusion

"The standard inverter provides a small area and is suitable for applications where high performance and flexible workload are required."

"For medium performance device that requires intermediate workload, staking and body biasing are feasible options for leakage reduction."

"DLS is the best leakage reduction technique but its main limitation is that it does not operate at a high frequency greater than 1 kHz in our used technology."

"Although body biasing shows low leakage current results, it requires a bias generator that adds power overhead."

A Robust Multi-Bit Soft-Error Immune SRAM Cell for Low-Power Applications [23]

This is a machine-generated summary of:

Abbasian, Erfan; Sofimowloodi, Sobhan: A robust multi-bit soft-error immune SRAM cell for low-power applications [23]

Published in: Analog Integrated Circuits and Signal Processing (2023)

Link to original: https://doi.org/10.1007/s10470-023-02144-0

If you want to cite the papers, please refer to the original.

For technical reasons we could not place the page where the original quote is coming from.

Abstract-Summary

"Portable applications and battery-operated devices require highly reliable, stable, and low-power nanometer-sized embedded cache static random access memory (SRAM) cells."

"The conventional 6-transistor (6 T) SRAM cell and its variants suffer from malfunctioning during the read/write operations, and instability, and are vulnerable to the multi-bit soft-error rate at scaled technology node and low supply voltage (V_{DD})."

"This paper proposes an 12 T SRAM cell with reliable functioning and reduced multi-bit soft-errors appropriate for low-power portable applications."

"The presence of stack transistors in the cell core and read path, and also high virtual ground (V_{GND}) minimize the leakage power dissipation."

"This cell shows at least 1.17×, 1.32×, and 1.04× smaller spread in read delay, RSNM, and WSNM, respectively, when subjected to PVT variations."

"The proposed cell eliminates the write half-select disturbance by employing a separate gate to drive the access transistors and thus column-interleaving structure and error correction coding can be applied to reduce multiple-bit upset and increase soft-error immunity."

Introduction

"Although an SRAM cell can dissipate lower power in the near-/sub-threshold region (where the V_{DD} is slightly higher/lower than the threshold voltage (V_{th}) of the transistor), it must face the increasing manufacturing process, voltage, and temperature (PVT) variations, reduced cell stability, degraded voltage margin, and prevailing leakage current in this region [24, 25]."

"In the advanced technology nodes coupled with V_{DD} reduction, the random dopant fluctuations (RDF) increase the V_{th} variations and lead to on-chip SRAM malfunctioning during the read and write operations because there is an exponentional relationship between SRAM static noise margin (SNM) and V_{th} in the sub-V_{th} region [25, 26]."

"For conventional 6-transistor SRAM cell (hereafter-called C6T), read SNM (RSNM) and write SNM (WSNM) are two main conflicting design metrics [27]."

"The C6T cell shows poor read stability at low V_{DD}, hence data stored in the cell may flip during the read operation, resulting in read upset [28]."

Related Works: Design and Challenge

"Two bit-interleaving SRAM cells were proposed in [29, 30] to enhance both RSNM and WSNM by utilizing isolated read path and data-dependent feedback cutting write-assist scheme, respectively."

"The proposed 12 T SRAM cell, that we hereafter call WFC12T SRAM cell, uses an inverter-based AND gate, offering full voltage swing in its output compared to an NMOS device used in SRAM cells designed in [27, 29, 30], to control write-access transistors and eliminate write half-select issues, reducing write failure probability under harsh PVT variations."

"The differential writing structure along with write data-dependent feedback-cutting write-assist technique used in the proposed WFC12T SRAM cell improves the WSNM and write delay."

"Our proposed design lacks the series-connected access transistors in its writing path and uses single-end bitline reading operation, which results in better improvements in the WSNM, write delay, and dynamic read power consumption, compared to 12 T SRAM cells presented in [31, 32]."

The Proposed WFC12T Cell and Its Operations

"During hold operation or standby mode, the WWL and CSL are kept at V_{DD} and GND, respectively, to set LWWL at GND for turning off the access transistors, M7/M8."

"The WBL and WBLB are kept at V_{DD} to turn on the M2/M5."

"The RWL is set to V_{DD} and hence M9 is turned on, while V_{GND} is kept at GND to provide a large read current for discharging the RBL voltage."

"Write operation begins with the WWL and CSL being set to GND and V_{DD}, respectively, in order to set LWWL at V_{DD}."

"Depending on the data to be written to Q/QB, the WBL/WBLB is set at V_{DD}/GND or GND/V_{DD}."

"Consider the case in which the logical value of '0' is to be written to '0' storing node Q. In this case, WBL/WBLB is kept at GND/V_{DD} and hence M2/M5 is turned on/off."

Simulation Results and Discussion

"The Kim12T and proposed WFC12T SRAM cells show the best T_{WA} because the writing in these cells is deferential and there is only one access transistor in their write paths (as compared to WWL12T cell)."

"The SEDF9T and TG9T SRAM cell consume slightly lower $P_{leakage}$ than that of FD8T cell due to being single-ended nature and the presence of stacked transistors in their read and write paths."

"Other SRAM cells (FD8T, SEDF9T, Kim12T, proposed WFC12T, and WWL12T) use dual-bitlines structure, and thus their write power is the highest."

"Although, the FD8T and proposed WFC12T cells use an inverter-based AND gate to drive access transistors during the write operation, which is controlled by row-and column-based signals, however, the FD8T cell consumes slightly lower write power."

"The proposed WFC12T cell shows 1.16× lower write power than that of the SEDF9T cell and incurs at least 1.42× higher write power when compared with other studied SRAM cells."

Conclusion

"We proposed a reliable 12 transistor SRAM cell to reduce multi-bit soft-error, namely WFC12T. The proposed cell used single-bitline read decoupling technique and write data-dependent feedback cutting-aware differential writing structure."

"The proposed cell minimized the $P_{leakage}$ through stacked transistors employed in right/left half-cell and read access buffer."

"The proposed WFC12T cell offered the second-best read power and shows a high critical charge under high-energy space radiation."

"All these together maked the proposed WFC12T cell a good choice for reliable, stable, and low-power portable applications and battery-operated devices."

BitMAC: Bit-Serial Computation-Based Efficient Multiply-Accumulate Unit for DNN Accelerator [33]

This is a machine-generated summary of:

Chhajed, Harsh; Raut, Gopal; Dhakad, Narendra; Vishwakarma, Sudheer; Vishvakarma, Santosh Kumar: BitMAC: Bit-Serial Computation-Based Efficient Multiply-Accumulate Unit for DNN Accelerator [33]

Published in: Circuits, Systems, and Signal Processing (2022)

Link to original: https://doi.org/10.1007/s00034-021-01873-9

If you want to cite the papers, please refer to the original.

For technical reasons we could not place the page where the original quote is coming from.

Abstract-Summary

"Contemporary hardware implementations of deep neural networks face the burden of excess area requirement due to resource-intensive elements such as a multiplier."

"A semi-custom ASIC approach-based VLSI circuit design of the multiply-accumulate unit in a deep neural network faces the chip area limitation."

"An area and power-efficient architecture for the multiply-accumulate unit is imperative to down the burden of excess area requirement for digital design exploration."

"It shows improvement by 25.94% in area, 35.65% in power dissipation, and 14.30% in latency with respect to the state-of-the-art multiply-accumulate unit design."

"In order to save leakage power, we exploit the power-gated design for the proposed architecture."

"The used coarse-grain power-gating technique saves 52.79% leakage/static power with minimal area overhead."

Introduction

"The MAC computational unit architecture needs to be optimized to increase the physical performance parameter of the deep neural network in terms of the area, power, and delay of the overall circuit."

"We optimize the overall circuit design using the proposed bit-serial computing-based MAC architecture to address the trade-off between area, power, and delay."

"We have designed an 8-bit MAC unit using a bit-level processing technique to improve the overall physical parameters like area, power, and delay of the circuit, and the power gating scheme is used for further static power saving."

"An 8-bit multiply-accumulate unit is designed using the proposed power-gated bitcell and the impact on the MAC performance is analyzed for parameters like area, power, and delay."

"For an 8-bit precision MAC, bitcell-based architecture takes 25.94% less area, 14.30% less delay, 34.65% less static power than state-of-the-art at 45 nm."

Background and Related Works

"The Wallace tree logic-based algorithm is designed to perform the multiplication in [34, 35]."

"To improve the performance by compromising the neural network's output accuracy, researchers have designed an approximate algorithm for multiplication and addition used in the MAC computational unit in [36–40]."

"In [39], approximate array-based multiplier is designed to improve circuit delay and power consumption."

"Stripes [41] shows that by using bit-serial computing, the network's performance can be increased without any accuracy loss in the computation."

"Tartan [42] also uses bit-serial computing units to improve precision flexibility."

"To overcome all these issues at lower technology nodes, we designed an efficient bit processing element and further high precision architecture of multiply-accumulate unit using bitcell."

"The proposed architecture intends to optimize the bit-serial computing to address the high-precision resources hungry MAC computation serially, which is efficient in terms of area, power, and critical path delay."

Semi-custom Design Architecture for Power Gated Bitcell

"The XNOR operation is performed between input(x) and weight(w), and then the output of the XNOR gate is added with the carry input(Carry in) using half adder."

"We have used the coarse-grain power gating technique in which we have used a single set of only 4 MOS and delay elements for a complete MAC unit that contain multiple bitcells, and hence, there is not much area overhead."

"As the adder operation depends upon the XNOR output (y_{out}), a delay element is included in the path from the clock to the two MOS connected with the adder block."

"The XNOR logic gate dissipates more static power when it is ON."

"To decrease the static power dissipation for positive half-cycle XNOR gate is power gated, i.e., when the XNOR block is in idle (hold) state."

Proposed Bitcell-Based Efficient Design of Multiply-Accumulate Unit for Deep Neural Network

"Each bitcell takes the input bit, processes it with the weight stored in that particular bitcell, and gives the 1-bit carry output and 1-bit sum output."

"The eight bitcells produce an 8-bit output called a partial product for a single bit input."

"For 8-bit input, there will be eight such partial products."

"These partial products are left-shifted depending upon in which input bit iteration the partial product is generated."

"In the first iteration, the LSB of the input, i.e., '1'b1' is given to the four bitcells in which 1-bit weight is stored in each bitcell."

"For the first bit cell, the initial carry is the complement of the input bit (1'b1), i.e., '1'b0' and for the remaining three bitcells, the initial carry is the output carry of the previous bitcell."

"The four bitcells provide a 4-bit partial product (4'b0101) for the first input bit that is LSB."

Experiment

"The proposed architecture is presented in HDL using the Verilog hardware description language."

"Addressing ASIC design, the RTL for our proposed 8-bit precision MAC architecture is synthesized, and results obtained by Design Compiler-Synopsys [43]."

"The CMOS extracted design is simulated in Cadence-Virtuoso [44] and extracted performance-parameter at typical typical (TT) process corners and mismatch."

"Addressing to the technology scaling impact, the multiply-accumulate architecture is synthesized at technology nodes of 45 and 180 nm."

"The physical performance parameters are extracted for proposed MAC architecture with and without the power-gating technique and for the related work."

Results and Discussion

"We design a logic without power gating technique and performance parameters is observed for both 180 and 45 nm technology node."

"We observed 31.24% and 39.19% less static power in power gated bitcell as compared with non-power gated bitcell at 180 nm and 45 nm technology node, respectively."

"From the observation to the downturn of the increasing static power dissipation, we have taken advantage of the power gating (PG) technique in the bitcell unit and the MAC design implementation."

"The proposed 8-bit MAC architecture is compared with the various state-of-the-art architectures on the important performance parameters like area, static and dynamic power, and critical path delay."

"The advantage of using the coarse-grain PG technique is that the relative impact of area increase in 8-bit MAC is much smaller when compared with the single power gated bitcell."

"The proposed 8-bit precision BitMAC shows promising performance compared with the state-of-the-art architectures."

Conclusion

"This paper presents a semi-custom ASIC design approach for VLSI design architecture that can implement logic using digital design technique and CMOS design technique that can be better in terms of area, delay, and power."

"The semi-custom ASIC design-based proposed model is synthesized at both 180 and 45 nm technology node."

"The results show that the proposed architecture is the best choice among all architectures."

"Further, to address the lower technology node, the coarse-grain power gating technique is used in the architecture to save the static power dissipation."

Design of Low Power Multiplier with Less Area Using Quaternary Carry Increment Adder for New-Fangled Processors [45]

This is a machine-generated summary of:

Gavaskar, K.; Malathi, D.; Ravivarma, G.; Priyatharshan, P. S.; Rajeshwari, S.; Sanjay, B.: Design of Low Power Multiplier with Less Area Using Quaternary Carry Increment Adder for New-Fangled Processors [45]

Published in: Wireless Personal Communications (2022)

Link to original: https://doi.org/10.1007/s11277-022-10006-7

If you want to cite the papers, please refer to the original.

For technical reasons we could not place the page where the original quote is coming from.

Abstract-Summary

"Because multipliers are employed in almost all arithmetic operations, they consume the majority of time and power."

"The array multiplier is one of the most efficient multipliers since it is both quick and simple."

"The basic components of an array multiplier are adders."

"The multiplier's efficiency will improve if the adders are much more efficient."

"The prior study used Complementary Metal Oxide Semiconductor (CMOS) to construct a Quaternary Carry-Lookahead Adder (CLA) that substitutes the adder circuit in the array multiplier."

"As a solution, the Quaternary Carry Increment Adder (CIA) is proposed."

"The proposed adder consumes less power and occupies less area than the existing Quaternary CLA."

"In the array multiplier, the Quaternary CIA substitutes the Quaternary CLA."

"This helps to lower the multiplier's power and area consumption."

Introduction

"Adders makes the building blocks of the multipliers."

"Circuits designed using Complementary Metal Oxide Semiconductor (CMOS) transistor has low power consumption, high noise immunity and it is widely used in VLSI chips [46]."

"Many researchers, with evolution in technology have already tried and are still endeavouring to craft multiplier units which furnishes either greater operating performance, less power consumption, regularity of layout and consequently with small area or even combination of them in one multiplier which makes them appropriate for several increased speed, decreased power and compact VLSI implementation [47].”

“In digital combinational circuits, an array multiplier is used for multiplying two binary numbers by utilizing an array of full adder and half adder.”

“Adders are building blocks of the multipliers.”

“The objective of the proposed idea is to design a multiplier with quaternary carry increment adder by reducing the factors like power and area.”

Existing Methods

“Higher radix number system like Quaternary Number System can be used to perform carry free addition [48].”

“This problem in the binary circuits can be overcome by using the number system having higher radix than binary number system.”

“With help of the higher radix, the number of interconnections in the circuits can be reduced.”

“In different logic circuits, Quaternary inverter plays a major role.”

“The AND function causes the AND circuit’s output to be the truncated value of numerous inputs.”

“The NAND circuit with two quaternary inputs is implemented.”

“The normal Carry-Look ahead logic is used to design the Quaternary gates.”

“The same propagate and generate logic of binary Carry-Look ahead.”

“Carry-Look ahead generated in order to generate carry every other bit with the help of the input carry C_{in}.”

“This quaternary adder can be used to perform fast addition than binary CLA due to lesser number of interconnections.”

Proposed Work

“In the proposed system, the quaternary CLA replaced with the 8-bit quaternary Carry Increment Adder in order to minimize the power consumption and area of the existing system.”

“The additional input Carry (C_{in}) is used for the completion of the full adder.”

“The carry digit and two quaternary digits are sum up by this circuit.”

“In order to implement carry free addition, the carry in for the adder circuit restricted to values upto 1, that is, the carry can be either 0 or 1 as in binary number system.”

“Half adder circuits are used as incrementor circuit in the Carry Increment Adder sequentially.”

“Four bits of LSB are added with the help 4 bit Quaternary RCA which produces sum and carry.”

"Since ripple-carry adder is used in place of carry-look ahead adder, the power intake of the circuit is reduced [49]."

Results and Discussion

"The output for the first set of the input as follows 8 bit Multiplicand, A [7:0] = 00000000 8 bit Multiplicand, A [7:0] = 00000000 8 bit Multiplier, B [7:0] = 00000000 16 bit Product, M [15:0] = 0000000000000000 The output for the second set of the inputs are as follows 8 bit Multiplicand, A [7:0] = 11111111 8 bit Multiplier, B [7:0] = 11111111 16 bit Product, M [15:0] = 1111111000000001 The glitches can be seen in the outputs, this is due to parallel inputs given to the adders in the circuit."

"Output: 16 bit Product, M [15:0] = 1,111,111,000,000,001 The glitches can be seen in the outputs, this is due to parallel inputs given to the adders in the circuit."

"By replacing the Quaternary CLA with Quaternary CIA in the array connection the power consumption is reduced."

"The Quaternary CIA is made up of 2 Ripple-Carry Adder and an Increment circuit."

Conclusion

"We proposed a Carry Increment Adder by using QSD number system in order to replace Quaternary Carry-Look ahead Adder in array multiplier circuit."

"These analyses are also done for the existing Quaternary Carry-Look ahead Adder multiplier and the results are compared."

"Quaternary Carry Increment Adder has a lower power consumption and occupies lesser area than the existing method."

"This is because, the number of transistors present in the proposed method is lesser than the existing Quaternary Carry-Look ahead Adder multiplier."

Ultra-Low-Power and Fast Voltage Level Shifter Using Muller C-Cell for VLSI Systems [50]

This is a machine-generated summary of:

Rajendran, Selvakumar; Chakrapani, Arvind: Ultra-Low-Power and Fast Voltage Level Shifter Using Muller C-Cell for VLSI Systems [50]

Published in: Circuits, Systems, and Signal Processing (2022)

Link to original: https://doi.org/10.1007/s00034-022-02098-0

If you want to cite the papers, please refer to the original.

For technical reasons we could not place the page where the original quote is coming from.

Abstract-Summary

"Voltage scaling is the significant technique to attain power-efficient operation utilizing voltage level shifters."

"Level shifters (LSs) with high performance are vital interconnecting circuits to achieve compactness in the multiple-supply voltage low-power VLSI systems."

"This article presents a fast and ultra-low static power LS adopting Muller C-cell along with stacked inverter."

"The structure of Muller C-cell helps to achieve ultra-low standby power in addition to fast switching, as expected."

Introduction

"The brief operation of differential cascode voltage switch (DCVS), current mirror (CM) and Wilson CM-based level shifters are presented along with their limitations."

"Even though the DCVS LS consumes almost zero static power at standby mode, it is affected with contention current (CC) issue for the lower input voltages and thus it increases delay and energy consumption."

"The CM-based LSs are affected with the static current issue for the change of input from low-to-high."

"The WCMLS is also inferior with the contention current (CC) and reduces voltage swing (RVS) issues."

"An ultra-low standby power and fast voltage level shifter/converter utilizing Muller C-cell is proposed to reimburse the limitations of the existing CM-based LSs."

The Proposed Muller C-Cell-Based Level Shifter (MCLS)

"The devices MN2 & MN3 are controlled by the input signal (A) while MP2 and MP3 are controlled by signals n4 and n2, respectively."

"An inverter consisting of stacked devices MP5 & MP6 and MN5 & MN6 is connected at the node n1 and produces the output signal (Z)."

"The pass transistor is utilized in the pull-down network in order to achieve fast fall/rise transition (high-speed operation) along with negligible leakage current without the input inverter."

"The node voltage at n1 turns MN5 & MN6 ON to discharge the output node Z to ground and turns MP7 OFF to disable the Wilson CM."

"When input changes from high-to-low, MN1 is turned ON to discharge the node n2 which activates MP3 to charge up the node n1."

Simulation and Performance Analysis of MCLS

"The stacking of devices and structure of Muller C-cell along with WCM in the MCLS provide superior static power performance."

"The feasibility of the MCLS is verified by selecting width (W) of all devices as 7 times the minimum feature size of the technology (180 nm technology) and observed its level conversion functionality."

"The standby power is larger for best case due to consideration of fast NMOS and fast PMOS devices and it is lower at worst case due to slow NMOS devices in the MCLS."

"It is perceived that the MCLS has adequate conversion range from 410 mV to 1.8 V. Moreover, MCLS demonstrates the superior static power performance and consumes very low energy and delay while converting low input signal."

"It is evident from the transient results that MCLS performs level conversion with measured output delay of 10.7 ns."

Conclusion

"This article proposed the fast and ultra-low-power Muller C-cell-based LS which performs level shifting to high output voltage (1.8 V) from low input voltage (410 mV)."

"The proposed MCLS achieves high performance exhibiting output delay of 2.4 ns and energy of 127 fJ/transition with help of pass transistor and DTMOS with INWE-based device sizing in the pull-down network."

"The MCLS consumes very low static power of 47.7 pW in the idle mode, using Muller C-cell with stacked inverter and Wilson current source in the pull-up network."

References

1. Aishwarya N, Nayak A, Pal I, Kumar V, Islam A (2019) A novel CNFET based tunable memristor emulator. Microsyst Technol. https://doi.org/10.1007/s00542-019-04486-0
2. Yu DS, Liang Y, Chen H, Lu HHC (2013) Design of a practical memcapacitor emulator without grounded restriction. IEEE Trans Circ Syst II 60(4):207–211
3. Deng J, Wong HSP (2007) A compact SPICE model for carbon nanotube field effect transistors including nonidealities and application-part II: full device model and circuit benchmarking. IEEE Trans Electron Devices 54(12):3195–3205
4. Chua LO (1971) Memristor—the missing circuit element. IEEE Trans Circ Theory 18(5):507–519
5. Pal I, Islam A (2018) Circuit-level technique to design variation- and noise-aware reliable dynamic logic gates. IEEE Trans Device Mater Reliab 18(2):224–239
6. Pal I, Kumar V, Aishwarya N, Nayak A, Islam A (2019) A VDTA-based robust electronically tunable memristor emulator circuit. Analog Integr Circ Sig Process. https://doi.org/10.1007/s10470-019-01575-y
7. Strukov DB, Snider GS, Stewart DR, Williams RS (2008) The missing memristor found. Nature 453:80–83
8. Alharbi AG, Fouda ME, Khalifa ZJ, Chowdhury MH (2017) Electrical nonlinearity emulation technique for current-controlled memristive devices. IEEE Access 5:5399–5409
9. Elwakil AS, Fouda ME, Radwan AG (2013) A simple model of double-loop hysteresis behavior in memristive elements. IEEE Trans Circuits Syst II Express Briefs 60(8):487–491
10. Kim H, Sah MP, Yang C, Cho S, Chua LO (2012) Memristor emulator for memristor circuit applications. IEEE Trans Circ Syst I Reg Pap 59(10):2422–2431
11. Petrovic PB (2018) Floating incremental/decremental flux-controlled memristor emulator circuit based on single VDTA. Analog Integr Circ Sig Process 96(3):417–433
12. Sánchez-López C, Mendoza-López J, Carrasco-Aguilar MA, Muñiz-Montero C (2014) A floating analog memristor emulator circuit. IEEE Trans Circuits Syst II Express Briefs 61(5):309–313
13. Sánchez-López C, Carrasco-Aguilar MA, Muñiz-Montero C (2015) A 16 Hz–160 kHz memristor emulator circuit. AEU Int J Electron Commun 69(9):1208–1219
14. Cam ZG, Sedef H (2017) A new floating memristance simulator circuit based on second generation current conveyor. J Circ Syst Comput 26(2):1–15
15. Ranjan RK, Rani N, Pal R, Paul SK, Kanyal G (2017) Single CCTA based high frequency floating and grounded type of incremental/decremental memristor emulator and its application. Microelectron J 60:119–128
16. Sánchez-López C, Aguila-Cuapio LE (2017) A 860 kHz grounded memristor emulator circuit. AEU Int J Electron Commun 73:23–33
17. Yunus B, Abdullah Y, Firat K (2017) Memristor emulator with tunable characteristic and its experimental results. AEU Int J Electron Commun 81:99–104
18. Sozen H, Cam U (2016) Electronically tunable memristor emulator circuit. Analog Integr Circ Sig Process 89(3):655–663
19. Babacan Y, Kaçar F (2017) Floating memristor emulator with subthreshold region. Analog Integr Circ Sig Process 90(2):471–475
20. Gilbert B (1968) A precise four-quadrant multiplier with subnanosecond response. IEEE J Solid State Circ 3(4):365–373
21. Kilani D, Mohammad B, Alhawari M, Saleh H, Ismail M (2019) Analysis and characterization of leakage reduction methodologies for stacking, body biasing and DLS in 65 nm CMOS technology. Analog Integr Circ Sig Process. https://doi.org/10.1007/s10470-019-01520-z
22. Valentian A, Beigne E (2008) Automatic gate biasing of an SCCMOS power switch achieving maximum leakage reduction and lowering leakage current variability. IEEE J Solid State Circuits 43(7):1688–1698

23. Abbasian E, Sofimowloodi S (2023) A robust multi-bit soft-error immune SRAM cell for low-power applications. Analog Integr Circ Sig Process. https://doi.org/10.1007/s10470-023-02144-0
24. Ataei S, Stine JE, Guthaus MR (2016) A 64 kb differential single-port 12T SRAM design with a bit-interleaving scheme for low-voltage operation in 32 nm SOI CMOS. In: 2016 IEEE 34th international conference on computer design (ICCD). IEEE, pp 499–506
25. Sharma V, Gopal M, Singh P, Vishvakarma SK, Chouhan SS (2019) A robust, ultra low-power, data-dependent-power-supplied 11T SRAM cell with expanded read/write stabilities for internet-of-things applications. Analog Integr Circ Sig Process 98:331–346
26. Lorenzo R, Pailly R (2020) Single bit-line 11T SRAM cell for low power and improved stability. IET Comput Digit Tech 14:114–121
27. Pal S, Islam A (2016) 9-T SRAM cell for reliable ultralow-power applications and solving multibit soft-error issue. IEEE Trans Device Mater Reliab 16:172–182
28. Calhoun BH, Chandrakasan AP (2007) A 256-kb 65-nm sub-threshold SRAM design for ultra-low-voltage operation. IEEE J Solid State Circuits 42:680–688
29. Sharma V, Bisht P, Dalal A, Gopal M, Vishvakarma SK, Chouhan SS (2019) Half-select free bit-line sharing 12T SRAM with double-adjacent bits soft error correction and a reconfigurable FPGA for low-power applications. AEU Int J Electron Commun 104:10–22
30. Sharma V, Gupta N, Shah AP, Vishvakarma SK, Chouhan SS (2021) A reliable, multi-bit error tolerant 11T SRAM memory design for wireless sensor nodes. Analog Integr Circ Sig Process 107:339–352
31. Pal S, Bose S, Ki W-H, Islam A (2019) Half-select-free low-power dynamic loop-cutting write assist SRAM cell for space applications. IEEE Trans Electron Devices 67:80–89
32. Chiu Y-W, Hu Y-H, Tu M-H, Zhao J-K, Chu Y-H, Jou S-J et al (2014) 40 nm bit-interleaving 12T subthreshold SRAM with data-aware write-assist. IEEE Trans Circuits Syst I Regul Pap 61:2578–2585
33. Chhajed H, Raut G, Dhakad N, Vishwakarma S, Vishvakarma SK (2022) BitMAC: bit-serial computation-based efficient multiply-accumulate unit for DNN accelerator. Circuits Syst Signal Process. https://doi.org/10.1007/s00034-021-01873-9
34. Janveja M, Niranjan V (2017) High performance Wallace tree multiplier using improved adder. ICTACT J Microelectron 3(01):370–374
35. Kesava RBS, Rao BL, Sindhuri KB, Kumar NU (2016) Low power and area efficient Wallace tree multiplier using carry select adder with binary to excess-1 converter. In: 2016 Conference on Advances in Signal Processing (CASP). IEEE, pp 248–253
36. Abed S, Khalil Y, Modhaffar M, Ahmad I (2018) High-performance low-power approximate Wallace tree multiplier. Int J Circuit Theory Appl 46(12):2334–2348
37. Jiang H, Liu C, Lombardi F, Han J (2018) Low-power approximate unsigned multipliers with configurable error recovery. IEEE Trans Circuits Syst I Regul Pap 66(1):189–202
38. Raut G, Bhartiy V, Rajput G, Khan S, Beohar A, Vishvakarma SK (2019) Efficient low-precision cordic algorithm for hardware implementation of artificial neural network. In: International Symposium on VLSI Design and Test. Springer, Singapore, pp 321–333
39. Sato T, Ukezono T (2020) A dynamically configurable approximate array multiplier with exact mode. In: 2020 5th International Conference on Computer and Communication Systems (ICCCS). IEEE, pp 917–921
40. Van Toan N, Lee JG (2020) FPGA-based multi-level approximate multipliers for high-performance error-resilient applications. IEEE Access 8:25481–25497
41. Judd P, Albericio J, Hetherington T, Aamodt TM, Moshovos A (2016) Stripes: bit-serial deep neural network computing. In: 2016 49th Annual IEEE/ACM International Symposium on Microarchitecture (MICRO). IEEE, pp 1–12
42. Delmas A, Sharify S, Judd P, Moshovos A (2017) Tartan: accelerating fully-connected and convolutional layers in deep learning networks by exploiting numerical precision variability. arXiv:1707.09068

43. https://www.synopsys.com/implementation-and-signoff/rtl-synthesis-test/design-compiler-graphical.html
44. https://www.cadence.com/en_US/home/tools/custom-ic-analog-rf-design/circuit-design/virtuoso-schematic-editor.html
45. Gavaskar K, Malathi D, Ravivarma G, Priyatharshan PS, Rajeshwari S, Sanjay B (2022) Design of low power multiplier with less area using quaternary carry increment adder for new-fangled processors. Wirel Pers Commun. https://doi.org/10.1007/s11277-022-10006-7
46. Kim S, Kang Y, Baek S, Choi Y, Kang S (2021) Low-power ternary multiplication using approximate computing. IEEE Trans Circuits Syst II Express Briefs 68(8):2947–2951
47. Antelo E, Montuschi P, Nannarelli A (2016) Improved 64-bit radix-16 booth multiplier based on partial product array height reduction. IEEE Trans Circuits Syst I Regul Pap 64(2):409–418
48. Kabra NK, Patel ZM (2021) A radix-8 modulo 2n multiplier using area and power-optimized hard multiple generator. IET Comput Digit Tech 15(1):36–55
49. Lee J, Seo H, Seok H, Kim Y (2021) A novel approximate adder design using error reduced carry prediction and constant truncation. IEEE Access 9:119939–119953
50. Rajendran S, Chakrapani A (2022) Ultra-low-power and fast voltage level shifter using muller C-cell for VLSI systems. Circuits Syst Sig Process. https://doi.org/10.1007/s00034-022-02098-0

Chapter 8
Challenges in Full Chip Optimization

Introduction by the Editor

This chapter discusses a lot of practices that can help minimise, if not remove all the challenges in full chip optimizations. This chapter talks about techniques Which can help a multi objective redundancy hardening task with efficient mapping for independent task on multiple cores. Chapter also walks through Methods of a multi objective learning automation, Techniques that can assistant floor planning and design optimal routing in VLSI circuits using particle swarm optimization, Floor planning of 3D IC design using hybrid optimiser. A novel hybrid method of a chimp Yes. with cuckoo search algorithm for optimal designing of infinite impulse response Filter using high level synthesis. Chapter elaborates on FPGA based hardware accelerator design, also a low-cost BIST design supporting automated testing of fabricated chips. These techniques can help increase efficiency and minimise challenges faced during full chip optimization.

This chapter overviews redundancy hardening with multiple objectives with efficient task mapping for scheduled tasks on multiple cores. This aims at achieving a low-cost and a high-performance system with task hardening. It's important to acknowledge min-max weight matching problem which is minimized. Study on learning automation with multiple objectives and its application on VLSI design is thoroughly mentioned in this chapter. This chapter also includes two indicators of power-delay product and power area product. These indicators are used extensively for optimization of all three critical factors in VLSI which are area, power and delay. This also highlights that the learning automation provide a new framework for circuit design. This chapter shows comparison data below traditional and proposed learning framework.

Adaptive Particle swarm optimization is an innovative method applied in VLSI floor planning is discussed in detail. It includes problem model formation of floorplan and development of ISI-APSO based floorplan model. Simulation results of the

A. S. Shastri (ed.), *Optimization Methods in VLSI Design*,
https://doi.org/10.1007/978-981-95-2457-0_8

mentioned technique are compared with conventional methods. FPGA based hardware accelerated design for real-time speech processing causes a lot of repetitive computes for big range for speech data. Hardware accelerated design helps with high hardware usage and reduced computation time. This highlights that linear prediction analysis provides efficient solutions. Floorplan for 3D IC design using multi-verse optimizer is reviewed in this chapter. It is used to optimize planar coordinates and dimension of nodes needed for optimal floorplan. One of the studies includes mathematical modelling of system identification problem, Chimp optimizer, Cuckoo search algorithm. All the techniques are employed to solve problems in HLS of IIR filters.

Optimal parameter prediction and evaluation of IIR filters for system identification is mentioned in this study which helps address model identification challenges

Machine Generated Summaries

Disclaimer: The summaries in this chapter were generated from Springer Nature publications using extractive AI auto-summarization: An extraction-based summarizer aims to identify the most important sentences of a text using an algorithm and uses those original sentences to create the auto-summary (unlike generative AI). As the constituted sentences are machine selected, they may not fully reflect the body of the work, so we strongly advise that the original content is read and cited. The auto generated summaries were curated by the editor to meet Springer Nature publication standards. To cite this content, please refer to the original papers.

Machine generated keywords: floor, floor plan, planning, floor planning, plan, multiobjective, speech, type, redundancy, bist, filter, hardware, select, transient fault, module

Multi-Objective Redundancy Hardening with Optimal Task Mapping for Independent Tasks on Multi-Cores [1]

This is a machine-generated summary of:

Yuan, Bo; Li, Bin; Chen, Huanhuan; Zeng, Zhigang; Yao, Xin: Multi-objective redundancy hardening with optimal task mapping for independent tasks on multi-cores [1]

Published in: Soft Computing (2019)

Link to original: https://doi.org/10.1007/s00500-019-03937-0

If you want to cite the papers, please refer to the original.

For technical reasons we could not place the page where the original quote is coming from.

Abstract-Summary

"As DMR and TMR have different time and cost overheads, we must carefully determine which one should be used for each task (i.e., task hardening) to achieve the optimal system design."

"For multi-core systems, the system-level design includes the allocation of cores for the tasks (i.e., task mapping) as well."

"This paper aims at task hardening and mapping simultaneously for independent tasks on multi-cores with heterogeneous performances, in order to minimize the maximum completion time of all tasks (i.e., makespan)."

"We demonstrate that once task hardening is given, task mapping of independent tasks can be achieved by employing min–max-weight perfect matching with a polynomial time complexity."

"Besides, as there is a trade-off between cost and time performance, we propose a multi-objective memetic algorithm (MOMA)-based task hardening method to obtain a set of solutions with different numbers of cores (i.e., costs), so the designer can choose different solutions according to different requirements."

"Our experimental studies have demonstrated the effectiveness of the proposed method and have shown that by combining the results of MOMA and MOEA we can provide a designer with a highly accurate set of solutions within a reasonable amount of time."

Introduction

"Space redundancy (i.e., replication), based on executing redundant replicas of task independently on different cores, does not require any specific fault detection mechanism and uses result comparison (majority voting) for fault detection and masking [2–4]."

"Both DMR- and TMR-based hardening techniques are considered, and we show how to link the problem of task mapping with the goal of minimizing the worst-case makespan (i.e., the maximum completion time of all tasks) to the MMW-PM model, and then, based on binary search and Hungarian algorithm [5], we use an efficient heuristic algorithm [proposed in our previous work [6] to obtain an MMW-PM from a bipartite graph with polynomial time complexity."

"Besides, as there is a trade-off between cost and time performance [7], we propose a multi-objective memetic algorithm (MOMA)-based task hardening method to obtain a set of solutions with different numbers of cores (i.e., costs), so the designer can choose a solution from the Pareto front according to user preferences, such as cost budget or performance requirement."

Problem Definition

"Assume we are given a resource library $C = \{c_1, c_2, \ldots, c_M\}$ with different frequencies $f_{max} = f_1 \geq f_2 \ldots \geq f_M = f_{min}$, and an instance consisting of independent tasks $T = \{t_1, t_2, \ldots, t_N\}$ with different numbers of instructions $d_{max} = d_1 \geq d_2 \ldots \geq d_N = d_{min}$, the goal of this paper is to (1) determine TMR mode or DMR mode for each task, and (2) allocate a core for each task (including all original tasks and new tasks), such that (1) the number of cores used, i.e., $K = 2K_{DMR} + 3K_{TMR}$, is minimized, where K_{DMR} and K_{TMR} are the number of tasks hardened by DMR and TMR, respectively, and $K_{DMR} + K_{TMR} = N$, and (2) the worst-case makespan, i.e., $wcet(T) = \max\{wcet(t_1), wcet(t_2), \ldots, wcet(t_N)\}$, is minimized."

"A naive way to find an MMW-PM (i.e., the optimal mapping) is to find all perfect matchings in the given bipartite graph first and then select the one whose maximal edge weight is minimal."

Multi-Objective Redundancy Hardening

"Each position of the vector describes the fault-tolerant technique that assigned to the task, i.e., $th_i = 0$ indicates that task t_i is hardened by DMR with re-execution and $th_i = 1$ indicates that task t_i is hardened by TMR. Crossover In GA, crossover is a genetic operator that recombines more than one (generally two) parent chromosomes to produce one or two child chromosomes from them."

"The applied fault-tolerant techniques are mixed for the task hardening solutions. Mutation In GA, mutation is a genetic operator that alters one or more gene values in a chromosome from its initial state to maintain the genetic diversity of the population."

"In our implementation, selection for reproduction is performed before the crossover operator is applied, which is based on a purely random basis without bias to filter any individual, and selection for survival is performed according to NSGA-II, i.e., a fast nondominated sorting approach and a novel selection process with the consideration of both fitness and spread."

Simulations and Discussion

"It can be seen that: (1) for small-scale problem (e.g., N = 10), MOEA-based hardening performs better than MOMA-based hardening, and (2) for large-scale problems (e.g., N = 30, 50 and 100), the mean IGD values obtained by MOMA-based hardening are much lower than those obtained by MOEA-based hardening on most test instances, which is an indication of better approximation to the Pareto optimal front."

"Another observation is that the results obtained by MOEA-based hardening are all partiality located at the area where $Object_1$ (number of cores) is minimized, especially for large-scale problems (e.g., N = 50 and 100)."

"As the time complexity ($O(n^3 log^n)$) of the Hungarian-based optimal MMW-PM method (Algorithm 2) is high, the whole EA process is very time-consuming for large-scale problems, e.g., more than 2 days for n = 100 test instances."

Conclusion

"This paper aims at jointly task hardening and task mapping for independent tasks on heterogeneous multi-core systems in order to achieve low cost and high performance."

"In order to minimize the worst-case makespan with optimal task mapping, we show that the mapping problem can be modeled as a min–max-weight perfect matching problem."

"As there is a trade-off between the cost and performance in task hardening, we model the task hardening problem as a multi-objective optimization problem."

"Since a MOEA can find a partial Pareto front only, we propose a problem-specific local search operator and incorporate it in the MOEA framework."

An Improved Multi-Objective Learning Automata and Its Application in VLSI Circuit Design [8]

This is a machine-generated summary of:

Sayyadi Shahraki, Najmeh; Zahiri, Seyed Hamid: An improved multi-objective learning automata and its application in VLSI circuit design [8]

Published in: Memetic Computing (2020)

Link to original: https://doi.org/10.1007/s12293-020-00303-8

If you want to cite the papers, please refer to the original.

For technical reasons we could not place the page where the original quote is coming from.

Abstract-Summary

"An improved multi-objective optimization method, based on learning automata (called IMOLA), is proposed and its performance on the design of a variety of functional circuits is investigated."

"To evaluate the capability of the proposed method on multi-objective problems, digital and analog circuits have been selected."

"Evaluation of the results was reported in terms of power-delay-product, power-area-product, success rate, Pareto-front, multi-objective criteria, circuit variables, design constraints, runtime, and performance analysis."

Introduction

"Reinforcement learning (RL) is one of the methods that can be effective in optimization and it is placed along with evolutionary approaches."

"In an RL system, the aim is to learn an optimal action by interacting with an uncertain environment."

"There is a feedback architecture in a learning system based on RL and the interaction between the learning agent and its environment can be described by a sequence of states, actions, and rewards [9]."

"Evolutionary algorithms are very effective for MO problems and various types of them are presented in the literature."

"By learning to choose the optimal action, the automata adapt themselves to the environment, needless to have detailed information about the environment model [10]."

"At first, a novel multi-objective learning automata has been introduced and called IMOLA for engineering problems."

Literature Review

"A new LA based selection hyper-heuristic controlling a set of MO metaheuristics introduced in [11]."

"The idea of using the Voting Q-learning was expressed by Tozer [12] and added a new optimization tool to the set of MOO methods."

"The voting methods used based on scoring and the Condorcet method as a component of a model-free many objective RL algorithm to allow agents to make globally Pareto optimal decisions with many conflicting objectives."

"In [13], an MO algorithm introduced using the clustering method."

"In [14], an algorithm implemented based on the diversity ranking so that established the robustness of the method in the form of convergence and diversity of the Pareto-fronts."

"In [15], a learning-based evolutionary MO algorithm introduced in which the LA method is used to select the appropriate mutation operator in the algorithm according to the feedback signal."

Proposed Approach: Circuit-Based Multi-Objective Optimization by Using the IMOLA

"There are 14 and 13 design parameters in digital and analog circuits, respectively."

"Design parameters in this circuit include width and length of transistors, biasing current (I_{bias}), compensation capacitance (C_c), and load capacitance (C_L)."

"In this amplifier, the biasing current, the total values of width and length of transistors are considered as design parameters."

"Two indicators of power-delay-product (PDP) for digital circuit and power-area-product (PAP) for analog circuit are used to evaluate the performance of the proposed methods in circuit optimization."

"An optimization tool is proposed for automated design of analog and digital circuits."

"The HSPICE simulates the circuit according to the received input and stores the output values including objective functions and design specifications in the HSPICE output (i.e., .lis file)."

"The proposed tool creates new design parameters for the next iteration by running the optimization process of two versions of MOLA and other algorithms."

Simulation Results

"The allowed ranges of the design parameters are assumed as follows: for level shifter circuit width of transistors is allowed to vary between [0.4, 1.6] μm, for two-stage CMOS op-amp circuit width of transistors is allowed to vary between [5, 40] μm, length of transistors can be changed between [0.25, 2] μm, the I_{bias} is allowed to vary between [20, 40] μA, the C_C is allowed to vary between [7, 14] pF, and the C_L is allowed to vary between [15, 16] pF, and for single-ended folded-cascode circuit width of transistors is allowed to vary between [0.24, 200] μm, length of transistors can be changed between [0.18, 5] μm, the I_{bias} is allowed to vary between

[30, 400] μA. In order to demonstrate the ability of the RL method to solve the problem of circuit optimization, the results are compared with several intelligent methods."

"The IMOLA for level shifter, two-stage CMOS op-amp, and single-ended folded-cascode circuits has the best SRs with values of 100%, 40% and 85%, respectively."

Conclusion

"LA performance as an intensive learning method for MO optimizing in the design process of VLSI circuits is evaluated and validated."

"The paper findings indicate that using LA methodology and especially its MO optimizer can provide a new framework for designing a variety of circuits, which can be a new approach for other research."

"The methodology of the proposed method is quite different from other conventional evolutionary methods used in circuit design and is based on LA, and the results also show the superiority of the learning automata-based method over circuit design algorithms."

"In the MO approach based on LA, as with other evolutionary algorithms, the most important issue is the proper adjustment of exploration and exploitation."

"In advanced circuit designs using the IMOLA method, a design framework is utilized to optimize the type of structure along with circuit sizing by algorithms."

Optimal Floor Planning in VLSI Using Improved Adaptive Particle Swarm Optimization [16]

This is a machine-generated summary of:

Vinay Kumar, S. B.; Rao, P. V.; Singh, Manoj Kumar: Optimal floor planning in VLSI using improved adaptive particle swarm optimization [16]

Published in: Evolutionary Intelligence (2019)

Link to original: https://doi.org/10.1007/s12065-019-00256-z

If you want to cite the papers, please refer to the original.

For technical reasons we could not place the page where the original quote is coming from.

Abstract-Summary

"Floor planning is necessary to design the VLSI circuit."

"A new Interactive Self-Improvement based Adaptive Particle Swarm Optimization (ISI-APSO) technique is proposed to enhance the exploration efficiency and accuracy than convolutional PSO."

"Within less computation time the proposed ISI-APSO technique attains best global search throughout the space."

"The simulation results show that the proposed ISI-APSO algorithm achieves better performance than other heuristic algorithms in exploring efficiency and speed of convergence."

"The implemented results demonstrate the importance of the proposed algorithm with respect to the robust performance."

Introduction

"Researchers have observed the advantages involved in VLSI floor planning [17–19] which includes numerous chips in it."

"A floor plan is decided depending upon the designing area, and it considers several kinds of modules, e.g., macros employed to model memory, design, Intellectual property (IP) cores and their requirements of space."

"The main motive of the floor planning technology is to reduce the cost parameters, e.g., length of wire, area of floor plan, etc so that the performance can be enhanced with reduced cost, less space, and reduced wire length."

"There exist two kinds of the floor plan. (1) Slicing the plan of the floor, i.e., the floor plan is divided to form two parts in vertical and horizontal lines. (2) Non-slicing of the floor plan, where there is no any partitioning of the modules [20]."

Literature Review

"Constraint and computational parameters adopted for designing the chips such as timing constraints, electrical characteristics, chip size, etc, are important as planning the floor is the main concept to optimize the objectives."

"Genetic algorithm depending on the normalized postfix encoding strategy has been presented in [17] to find a solution for floor planning technique."

"O-tree based specification has been addressed in [21] for solving the issues occurring in VLSI floor planning depending on GA."

"Ant colony optimization has been implemented in [22] for reducing the floor planning areas including the constraints of cluster."

"Floor planning including a 2.5-D layout solution has been adopted in [23] to minimize the interconnected length of wire."

"Modeling a test site chips in macros with rectangular shapes without wire and overlaps has been proposed in [24] for floor planning in VLSI design."

Problem Model of Floor Planning in VLSI

"The target of the floor planning is to offer a certain space for all these modules at reduced cost and without any overlap between them which can be considered as the area of floor plan and entire length of internal wires."

"The entire length of wire in planning the floor can be calculated by considering the required wire for each of the connections."

"The entire area covered by the floor plan is calculated as the small area of a rectangle, which is covered in each of the blocks."

"The objective function can be defined as the mathematical relation among the interested objects whose cost consumption has to be reduced."

State-of-the-Art of PSO Algorithms

"The large weight of inertia offers global solution whereas, a small weight in inertia allows for integrating local areas."

"Computation of PSO is based on the value of variables that are chosen, but there is no chance to measure the accurate values for these variables, hence the better way is to improve the adapted surrounding so that depending on the necessity, it can cope up with the suited value."

"The larger inertia value at the starting and smaller value at the final stage are said to be the better values that could adjust between the local and global searches."

"If there exists any information about the current population diversity, related variables in searching operation can allot better optimal values."

Proposed ISI-APSO-Based Floor Planning Model

"It is important to explain the position of particle in each iteration, when inertia weight scheme is adopted within the particle."

"This function offers required optimal solution."

"The inertia weight is updated for each successive iteration till a desired optimal solution is reached."

Simulation Results

"In test case 2, when worst performances are evaluated, the results show that the proposed method is better than PSO by 51%, better than ACO by 30%, better than SA by 50% and better than DE by 31%."

"For test case 2, the best performance of proposed ISI-APSO method is better than GA by 52%, better than PSO by 38%, better than ACO by 61%, better than SA by 48% and better than DE by 57%."

"For test case2, at population size 40, the proposed technique is superior to PSO by 22%, superior to ACO by 13%, superior to SA by 21% and superior to DE by 23%."

"From the results of simulation, it can be proved that the proposed technique offers better wire length when compared with other methods."

Conclusion

"From the analysis, it was shown that the proposed method offers better efficiency in accuracy and exploration than the traditional methods."

"The results achieved from simulation have shown that the performance of the proposed ISI-APSO algorithm provides better computation than other heuristic algorithms on considering the exploring efficiency and speed of convergence."

"When compared with the conventional methods, the proposed algorithm was found to be 70% better than GA, 96% better than PSO, 83% better than ACO, 91% better than SA and 83% better than DE."

"The performance of the proposed method was proved to offer better results than other conventional methods on solving floor planning problems in VLSI."

FPGA-Based Hardware-Accelerated Design of Linear Prediction Analysis for Real-Time Speech Signal [25]

This is a machine-generated summary of:

Singh, Dilip; Chandel, Rajeevan: FPGA-Based Hardware-Accelerated Design of Linear Prediction Analysis for Real-Time Speech Signal [25]

Published in: Arabian Journal for Science and Engineering (2023)

Link to original: https://doi.org/10.1007/s13369-023-07926-2

If you want to cite the papers, please refer to the original.

For technical reasons we could not place the page where the original quote is coming from.

Abstract-Summary

"This technique involves repetitive computations on a wide range of incoming audio data, leading to high resource consumption and execution time."

"The study focuses on optimizing linear prediction analysis at the sub-block level, specifically by modifying the autocorrelation and Levinson–Durbin algorithm to enhance the performance of the overall system."

"The suggested algorithm is integrated into a hardware/software co-design as a high-performance intellectual property with an AXI4-Stream interface."

"These findings are significant for designing optimized hardware for low bit rate speech coders with improved execution time and reduced resource consumption."

Introduction

"Linear prediction coding is utilized for both speech coding algorithms and recognition systems [26]."

"The linear prediction analysis is an estimation process for determining the LPC parameters of specific signal data."

"Among such approaches to improve system performance, FPGA-based hardware acceleration is the most prevalent for multimedia applications owing to reconfigurable resources and enhanced flexibility [27]."

"An FPGA-based hardware accelerator is designed for linear prediction analysis IP employing high-performance AXI4-Stream and AXI4-Lite interfaces."

"The main contributions of the present paper are as follows: A novel design and implementation of the optimized linear prediction analysis block in Xilinx Vivado (Verilog), Xilinx HLS (C/C + +), and MATLAB are carried out."

"A HW/SW co-design solution is proposed for the linear prediction analysis in Xilinx SDK (C/C + +)."

"Design of hardware accelerator and real-time simulation on Zynq Zybo FPGA board."

Background and Related Work

"An FPGA-based design for a real-time image denoising filter is implemented on Nexys-4 DDR for hardware acceleration in [28]."

"The following main performance metrics of the algorithm shall see a considerable improvement as a result of the creation of the dedicated hardware for the LPA: Performance Computing LPA coefficients can be expensive when dealing with large data sets or high-order prediction models."

"The LPA block is designed and implemented over FPGA in such a way that it requires limited resources compared to the reported works and enhances the overall performance of the architecture through hardware acceleration."

"To the best of the authors' knowledge, no prior works reported in the literature propose FPGA-based hardware acceleration in the field of low bit rate speech coding applications, where LPA is a crucial block for the evaluation of linear prediction coefficients."

Linear Prediction

"These are computed in multiple stages, including filtering and windowing, analysis of the speech segment for autocorrelation estimates, and LPC estimation for LP parameters."

"Due to the non-stationary nature of speech waveforms, speech signals are often divided into finite-length frames before being transmitted to the actual analysis and coefficient estimation."

"As the signal is processed frame by frame, framing contributes to the actual implementation of speech coding."

"Autocorrelation plays a vital role in linear prediction analysis where linear prediction parameters are calculated for the particular signal frame."

"The recursive approach uses shorter frames of nearly 20 samples, each with infinite-length windows such as the Chen or Barnell window, which increases the hardware's computational load but minimizes the speech coding delay."

"The hardware accelerator must be closely linked with the main CPU to enable rapid data transfer."

Hardware Design of the LPA

"The resources required for the computation of output coefficients are significantly higher, which are eventually designed and tuned to reduce hardware complexity."

"The autocorrelation estimation block requires the multiplication of two samples at the same time and addition with the past samples to obtain one autocorrelation coefficient."

"In state 'S3,' the Levinson–Durbin block starts processing the data obtained in state 'S2' by storing two of the autocorrelation coefficients into temporary 32-bit registers. '"

"The data that are being processed in the autocorrelation estimation are of integer data type because there was only multiplication involved for the autocorrelation estimation, but when these outputs are fed to the LD block, they pass through division, which results in a fractional output."

"The output of the autocorrelation block is a signed 32-bit integer that is fed to the Levinson–Durbin (LD) block for LPC coefficient estimation."

Results and Discussion

"For shift register-based design (P1), the consumption of FFs and LUTs is reduced to almost 50%, and a reduction in latency of nearly 11% is observed compared to the work reported in [29]."

"When comparing the slices consumption, the proposed design P1 shows a minimum reduction of 26.64% and a maximum reduction of 97.12% than the works demonstrated by [30] and [31], which makes the proposed design P1 more optimized in terms of area or resource consumption."

"The proposed design P2 performs similarly compared to the P1 but shows an improvement in FFs consumption by 11.26%."

"The proposed design P1 shows a maximum improvement of 89.76% and a minimum improvement of 14% in clock frequency when compared with the LPA architectures reported in [31] and [30]."

Conclusion

"Work, a custom IP block for linear prediction analysis is developed, which can be used as an independent module in any speech coding algorithm designed in FPGA."

"The proposed IP block also improves the clock frequency by an average of 89% (maximum) and 10% (minimum) compared to the previous works reported in the literature."

"The validation results demonstrate that the proposed design achieves a speedup of over 99.56% compared to the CPU-simulated results when tested on a Zynq Zybo FPGA board using random speech input."

"The proposed LPA IP block offers a practical, effective, and efficient solution for speech coding in FPGA-based systems, benefiting from the significant resource and speed improvements achieved in the present work."

A Low-Cost BIST Design Supporting Offline and Online Tests [32]

This is a machine-generated summary of:

Menbari, Ahmad; Jahanirad, Hadi: A Low-cost BIST Design Supporting Offline and Online Tests [32]

Published in: Journal of Electronic Testing (2022)

Link to original: https://doi.org/10.1007/s10836-022-05986-0

If you want to cite the papers, please refer to the original.

For technical reasons we could not place the page where the original quote is coming from.

Abstract-Summary

"Offline and online built-in self-test (BIST) designs are low-cost platforms to test very complex modern chips."

"The offline BIST design embeds the test pattern generator (TPG) into the chip to be activated in the test time."

"The online (or concurrent) BIST design eliminates the TPG and utilizes the system's input vectors to accomplish the test process."

"This paper proposes a BIST design that supports both online and offline tests."

"In the online part of the design, a selector module passes the input vectors which belong to a pre-computed test set to the reduction part."

"The test set contains the test vectors, which generate 0 remainders in the division by the LFSR's polynomial of the selector."

"In the concurrent test latency (CTL) aware design, the size of the test set is expanded by adopting the selecting part to select the test vectors which generate the same remainders in the division by two different polynomials."

"The small size of the test set and the compacting test vectors resulted in a tremendous reduction of hardware overhead."

Introduction

"Built-in self-test (BIST) approaches are classified into offline and online categories [33]."

"In the offline-BIST design, the regular operation of the circuit under test (CUT) stalls, then a test pattern generator (TPG) applies test vectors to the CUT's inputs."

"The online BIST approach solves the above problems utilizing the applied input vectors to test the CUT."

"The continuous test process of the online BIST approach would increase the detection probability of intermittent and transient faults [34–37]."

"The online BIST design employs a selecting part to select the input vectors which belong to a pre-computed test set [38]."

"A significant drawback of the online BIST approach is its long test time completion."

"This paper proposes a BIST design that can perform both offline and online tests."

"A low-cost TPG is employed to test the CUT when offline test mode is selected."

Related Work

"The CTL of duplication design is minimum because all system input vectors belong to the test set."

"Some concurrent BIST designs are based on a pre-computed test set in which the test vectors are selected by deterministic TPG algorithm [39–44]."

"In the offline mode, a set of 2 × 1 multiplexers connect the generated test vectors by the t-stage counter and (n-t)-bits from an OR-plane to the CUT inputs."

"The multi-level decoder is implemented by 2-input AND gates wherein if one of the test vectors occurs in the input of CUT; the related output is set to 1."

"As a problem, the hardware complexity of decoder modules in both NEMO and CE-NEMO is highly dependent on the number of pre-computed test vectors."

"Similar to NEMO and CE-NEMO, the hardware overhead of the DC-based method is highly dependent on the number and bit-width of pre-determined test vectors."

Preliminaries

"An N-bit binary number is converted to an algebraic polynomial of order N − 1."

"$1 + 1 = 0$, $1 + 0 = 1$, $0 + 0 = 0$, $1 \times 1 = 1$, $1 \times 0 = 0$, $0 \times 0 = 1$. For example, the addition and multiplication of 110 and 101 are calculated as follows: $110 \Rightarrow X^2 + X$, $101 \Rightarrow X^2 + 1$, $110 + 101 = X^2 + X + X^2 + 1 = 011$, $110 \times 101 = (X^2 + X) \times (X^2 + 1) = 11{,}110$ All the LFSRs in our proposed BIST design are Type-2 LFSR (internal XOR LFSR)."

"The locations of inserted XOR gates are determined based on the LFSR's characteristic polynomial."

"If the corresponding polynomial of the N-bit binary number is G(x) and the characteristic polynomial of the LFSR is L(x), then the final content of the LFSR's DFFs would be the remainder of the division of G(x) by L(x)."

"There are 2^N G(x) polynomials of order N. If these polynomials are divided by L(x) with the order of k, the set of remainders would be generated as follows."

Proposed BIST Design

"In the concurrent test latency aware (CTL-aware) approach, the selected test vectors are defined as the input vectors which generate the same reminders in the division by two different polynomials, $L_1(x)$ and $L_2(x)$."

"Based on the number of D-ALG's test vectors (#test_vec$_{D\text{-}ALG}$), construct the L(X)'s coefficient polynomial set."

"The input vectors which generate an equal remainder in the division by $L_1(x)$, $L_2(x)$, and $L_3(x)$ would belong to the test set."

"The goal of the reduction part is to compress the width of the selected test vector and the CUT's output vector for simplification of GC design."

"The only constraint that should be satisfied is that the generated reminders for selected test vectors in the division by $L_{red}(x)$ should be different."

"In the CTL-aware approach, the selected test vectors are different from the generated test vectors by offline TPG."

The Detailed Design of C6288

"The selecting part is designed based on the HW-aware approach."

"$LFSR_1$ with L(x) characteristic polynomial is inserted in the selecting part."

"The offline part imposes a large hardware overhead."

"Deploying the serial approach would result in a 69% hardware overhead in the cost of more offline test time."

Simulation Results and Comparisons

"Columns, the hardware overheads of the online part in the HW-aware and the CTL-aware approaches are reported for 80% and 90% fault coverage values."

"The hardware overhead in the case of small circuits is 1000% on average which is due to a large equivalents gate metric of registers and XOR gates compared to the size of CUTs."

"For large and very large circuits wherein the LFSRs impact diminishes, the hardware overhead of the HW-aware approach is significantly smaller than DC-based design (approximately 10% on average)."

"The CTL-aware hardware overhead approximately equals the DC-based design."

"Except for limited cases such as s349 and C6288 circuits, the hardware overhead of CE-NEMO and NEMO are greater than both HW-aware, and CTL-aware approaches."

"Even for the large circuits in which the hardware overhead of MICSET reduces below 100%, the related hardware overhead of our BIST design would become less than 50%."

Conclusion

"A new BIST design has been proposed which supports both offline and online tests-designing a low hardware overhead TPG for the offline part along with an efficient method for compaction of test vectors bit width results in a very low hardware overhead (5% for very large circuit)."

"A CTL-aware test vector selection approach is proposed for which the test time reduces 100 times in average for benchmark circuits."

"The method is more efficient for large and very large circuits due to a smaller hardware overhead of LFSRs utilized to test vector selection and compression parts."

Floor Planning of 3D IC Design Using Hybrid Multi-Verse Optimizer [45]

This is a machine-generated summary of:

Pothiraj, Sivakumar; Kadambarajan, Jeya Prakash; Kadarkarai, Pandiaraj: Floor Planning of 3D IC Design Using Hybrid Multi-verse Optimizer [45]

Published in: Wireless Personal Communications (2021)

Link to original: https://doi.org/10.1007/s11277-021-08166-z

If you want to cite the papers, please refer to the original.

For technical reasons we could not place the page where the original quote is coming from.

Abstract-Summary

"Several research works have been undergone in the 3D IC floor planning concepts due to its higher demand and technological improvement."

"Floor planning is a complex step when there is a more number of a component on each layer of IC."

"In this proposed work, hybrid multi verse optimizer floor planning is utilized where the node coordinates and dimensions are adjusted to attain the optimized floor planning with minimal wire length and area using best universes."

Introduction

"When designing 3DIC, it is required to concern the thermal issues at every single stage and floor planning process as well [46]."

"A thermal driven floor planning algorithm was suggested by Cong in reference [47] utilizing simulated annealing based optimization technique which takes more run time."

"The overlapping of modules usually happens, here force directed floor planning algorithm has been utilized to resolve these issues."

"Reduced on chip temperature with smaller run time is achieved applying only lesser number of TSVs in floor planning stage."

"To resolve this issue [48], Guo presented O-tree representation based on non-slicing floor plan which attains minimum wire length and dead space."

"The search space size of floor planning problem for B tree representation is O (n!"

"B tree is efficient algorithm for floor planning."

Related Works

"A modified corner list algorithm named as novel 3D non-slicing floor planning representation was utilized and its properties were also derived."

"At the end, to obtain close optimal placement with lesser run time, the heuristic algorithm should be associated with MCL strategy which resolves the intricacy of floor planning [49]."

"In [50], as area minimization is the significant task in floor planning, author mainly focused on it to obtain optimal placement solution."

"Most of the researches have been done in floor planning in relation to area and wire length [51] but still there is a problem in acquiring solution for optimal placement."

“Because of non-deterministic polynomial-time hard problem in floor planning, hybrid Cuckoo search and particle swarm optimization have been utilized to resolve these concerns.”

“The subsequent problems were identified in existing methods: The optimization problem could be resolved through proper placement of TSV using floor planning and partitioning methods.”

Proposed Work

“After updating of the TSV only, the moth flame optimization will be formed, or else the new position of worst frog is calculated in every single sub-memeplexes.”

“The solution acquired from MVO is related to a universe (i.e., the variable prevailing in solution matches with the object of universe) in which the inflation rate equals fitness function of the solution and finally the term time relates to iteration.”

“The best universe can be learnt through shifting entire universes that hold objects using wormholes irrespective of inflation rate.”

“Each objects in the universe moves arbitrarily to the best universe irrespective of inflation rate.”

“Calculate fitness function utilizing the updated universes.”

“Compare it with the estimated inflation rate and update the best universe at each iteration.”

Performance Analysis

“The flexibility of algorithm for TSV optimization and floor planning methodology is also verified with a range of measures which are attained after the experimental approach.”

“Our proposed methodology attains better floor planning using hybrid multi verse optimizer.”

“The component placements are shown in the resulting floor plan and it was found that the proposed framework achieved better floor planning than the other existing approaches.”

“The number of TSV, WL and time are the parameters utilized to evaluate the performance of proposed methodology.”

“The proposed method achieved lesser time and TSV count when compared to the existing methodologies.”

“In circuit n100 for 350 TSV counts, our proposed method attained lesser WL range as 125,504 and time as 2.90.”

“The number of TSV, WL and time are the various parameters used to assess these performance of proposed methodology.”

Conclusion

"It is concluded that optimization is necessary to achieve effective TSV count."

"Several algorithms were surveyed to analyze the efficiency of optimization in TSV."

"After creating the TSV aware 3D IC partitioning system by reducing the layer components and giving rise to appropriate TSV count and fitness function based optimization reducing area and wire length somewhat, MVO was applied to optimize it further."

"Our hybrid multi verse optimizer floor planning achieved better wire length optimization than other approaches."

A Novel Hybrid of Chimp with Cuckoo Search Algorithm for the Optimal Designing of Digital Infinite Impulse Response Filter Using High-Level Synthesis [52]

This is a machine-generated summary of:

Kaur, Mandeep; Kaur, Ranjit; Singh, Narinder: A novel hybrid of chimp with cuckoo search algorithm for the optimal designing of digital infinite impulse response filter using high-level synthesis [52]

Published in: Soft Computing (2022)

Link to original: https://doi.org/10.1007/s00500-022-07410-3

If you want to cite the papers, please refer to the original.

For technical reasons we could not place the page where the original quote is coming from.

Abstract-Summary

"The use of infinite impulse response (IIR) models for identification is chosen over their equivalent finite impulse response (FIR) models, because the former yield more perfect models of physical plants for real-world optimization problems."

"Infinite impulse response (IIR) model structures tend to make various multi-modal error surfaces whose objective cost of the problems is significantly more difficult to minimize."

"A novel hybrid variant of chimp and cuckoo search method has been developed for IIR related large-scale optimization issues, and it is known as chimp–cuckoo search algorithm (ChCS)."

"The strength of the proposed method has been tested on the 23 standard functions and three types of infinite impulse response (IIR) models."

Introduction

"Regarding to finite impulse response (FIR) filters, Aggarwal et al. [53] proposed the cuckoo search algorithm (CSA) for filter design with best response in terms of band-stop attenuation and with the smallest band-pass ripple, with lower computational cost than genetic algorithm (GA) and particle swarm optimization (PSO)."

"Yadav and others Yadav et al. [54] applied grasshopper optimization algorithm (GOA) to design a linear phase finite impulse response (FIR) low-pass, high-pass, band-pass and band-stop filters and proposed methodology target to reach minimum absolute error difference fitness function, through selecting optimal filter coefficients."

"To develop proper IIR filter designing method as a multi-objective optimization problem, Wang, Yu, Bin Li, and Yunbi Chen Wang et al. [55] have proposed a new local search operator enhanced multi-objective evolutionary algorithm (LS-MOEA)."

Motive of This Study

"The aim of the current study is the application of a novel hybrid MHAs based on the combination of ChOA and CS called ChCS for the optimal parameter estimation of IIR filters for the best model identification and for HLS of data paths in digital filters."

"The proposed ChCS algorithm is tested on three different orders of IIR model."

"The main contributions of this paper are: A hybrid algorithm called ChCS that includes features from ChOA and CS is proposed."

"The ChCS performance is developed to solve the infinite impulse response (IIR)."

"Statistical and qualitative result analyses assess the performance of the proposed ChCS algorithm compared to other competitive algorithms."

The Mathematical Formulation of the System Identification Problem

"The IIR model or system identification is the most complicated optimizer issue."

"Model or system identification is the mathematical formulation of an unknown model or system by using only output and input data."

"Further, the mathematical formation of IIR model for system identification Cuevas et al. [56] has been illustrated through the following mathematical equations."

Chimp Optimizer Algorithm (ChoA or Chimp)

"This algorithm is inspired by individual intelligence and sexual motivation of chimps in their group hunting."

"During the searching of the global optima solution in the search domain, four different categories were applied for simulating diverse intelligence such as attacker, barrier, chaser and driver."

"This phase has applied the behavior of the chimps mathematically."

"In the next phase, the four best solutions yet obtained are stored and the rest of the search agents has forced to update their own positions according to the best search agents locations."

Cuckoo Search Algorithm (CS)

"This is a most robust nature-inspired algorithm; it was developed by Yang and Deb [57], Yang [58]."

"This strategy was originated based on the obligate brood parasitic behavior of some cuckoo species."

"It is extended by the so-called Lévy flights rather than by simple isotropic random walks."

"The pseudocode of CS algorithm is reported in Algorithm 2."

Hybrid ChCS Algorithm

"This strategy has been developed by merging the phases of the two robust population-based algorithms such as chimp and cuckoo search algorithms."

"In this development, the phases of cuckoo search algorithm play an important role for replacing and providing the best possible optima values during the searching of the target or goals in the complex domain."

"Part of this section have been illustrated the brief details of the proposed algorithm how it has worked for searching the target in the complex domain."

"These best fitness values of each member of the crowd are applied to search for the next best positions for trapping the goals in the complex domain."

"CS Phase: This phase plays an important role for giving the best optima values during the searching the goal and for a new positions."

"It helps to ignore the local optima for fast exploring the global best optima solutions for the chimp optimizer algorithm."

High-Level Synthesis (HLS) of the IIR Filters

"Of the work has implemented the ChCS algorithm for data paths in infinite impulse response (IIR) digital filters in the terms of delay, area and power, respectively."

"Robust metaheuristics have applied to reduce design time at lower levels and leading to superior circuit indices Kim et al. [59]."

"The HLS is the initial state in synthesizing a DFG (data flow graph) and circuit has applied to show the behavioral explanation, which describes the operators type and relationships amid them."

"In this graph, 1–4 outputs and 5–6 inputs operators are applied."

"7 input and 8 output operators are illustrated in this graph."

Implementation Stages of the Hybrid ChCS Algorithm on IIR

"The ChCS algorithm has been used for evaluating the chip area, delay and power in HLS of data paths in IIR."

"The obtained outputs have been verified against the outputs of the MFO [60] and PSO [60]."

"The implementation stages of the proposed algorithm have been described through the following steps: The position of the chimps or search agents plays an role of decision variables of the given function in the complex space."

"The each member modified their locations during the search process in the search space w.r.t."

Results Analysis of 23 Test Suites

"For testing and evaluating the robustness of the proposed ChCS algorithm, 23 standard benchmark test suites have been used."

"These suites have been used for tested the performance of the proposed evolutionary method and its performance compared with recent algorithms such as modified particle swarm optimization (MPSO) [61], sine–cosine algorithm (SCA) [62], time varying acceleration particle swarm optimization (TACPSO) [63], particle swarm optimization (PSO) [64] and chimp optimization algorithm [65]."

"The robustness of the algorithms have been verified through different categories of the test suites such as unimodal, multimodal and fixed-dimension multimodal."

"The robustness of the proposed method has been tested on 23 benchmark suites as unimodal, multimodal and fixed-dimension multimodal."

"We are verifying the solution accuracy of the evolutionary algorithm on 23 standard benchmark test suites."

"All experiments give strong evidence that the ChCS algorithm can be able to get the best optima solution fast during the search process than others."

Results Analysis on IIR Model Identification

"During the testing of this issue, the proposed algorithm has been implemented to recognize a second-order plant by a first-order infinite impulse response (IIR) for system identification."

"On the basis of obtained solutions we can observed that the proposed algorithm is able to find the best parameter values for this first-order IIR model for system identification."

"We have implemented the proposed algorithm on the second-order infinite impulse response (IIR) model for evaluating the identification of a second-order plant."

"We have implemented the proposed algorithm on the high-order infinite impulse response (IIR) model for evaluating the identification of a superior order plant."

"The proposed algorithm performs superior than CS, ABC, EM on first-order IIR and PSO, CS, ABC on second-order IIR models."

"It can be concluded that the ChCS algorithm is superior to PSO, ABC, EM, CS, FPA and ChoA algorithms in the high-order IIR model for system identification."

Results Analysis on IIR Digital Filter

"For different modes, the mean of the acquired response for a 50 times execution for the ChCS algorithm has been tabulated along with their relevant the standard deviation(sd)."

"The ChCS method is a superior than the MFO [60] and PSO [60] algorithms with average values 5599.7, 2945.09, 2823.99 and 6208.99, 3012.6, 3066.7 for the lowest power consumption and occupied area."

"The results 5, 6.88 and 6.54 of delay revealed that the ChCS algorithm is able to reduce the delay time as comparison with others for IIR digital filters issue."

"All simulations concluded that the ChCS algorithm is able to finds the superior quality of output responses for delay, area and power consumption for HLS of very large-scale integration (VLSI) circuits."

Conclusion and Future Work

"The infinite impulse response (IIR) model identification and HLS of data paths in digital filters issues are a big challenge for the researcher and engineers."

"The objectives of this study could be divided into two categories: (i) first to prove the effectiveness of the proposed algorithm, it has been utilized for optimization of HLS of data paths in IIR digital filters and (ii) secondly for identify the best optimal parameter estimation of IIR filters for the best model identification, so that can tackle model identification difficulties."

"ChCS is able to reduce the complexity of the HLS of data paths in IIR digital filters problem in the terms of lowest delay, area and power, respectively, as compared to others."

"The future work will be concentrated on two parts: (i) high-level synthesis of data paths, IIR model identification of different orders of model and (ii) developing novel multi-objective and hybrid metaheuristics for these tasks."

References

1. Yuan B, Li B, Chen H, Zeng Z, Yao X (2019) Multi-objective redundancy hardening with optimal task mapping for independent tasks on multi-cores. Soft Computing. https://doi.org/10.1007/s00500-019-03937-0
2. Koren I, Krishna CM (2007) Fault-tolerant systems. Morgan Kaufmann, San Francisco
3. Pradhan DK (1996) Fault-tolerant computer system design. Prentice-Hall, Upper Saddle River
4. Salehi M, Ejlali A, Al-Hashimi BM (2016a) Two-phase low-energy N-modular redundancy for hard real-time multi-core systems. IEEE Trans Parallel Distrib Syst 27(5):1497–1510
5. Kuhn HW (1955) The Hungarian method for the assignment problem. Naval Res Logist Q 2:83–97
6. Zhong F, Yuan B, Li B (2016) A hybrid evolutionary algorithm for multiobjective variation tolerant logic mapping on nanoscale crossbar architectures. Appl Soft Comput 38:955–966
7. Erbas C, Ceraverbas S, Pimentel AD (2006) Multiobjective optimization and evolutionary algorithms for the application mapping problem in multiprocessor system-on-chip design. IEEE Trans Evol Comput 10(3):358–374
8. Sayyadi Shahraki N, Zahiri SH (2020) An improved multi-objective learning automata and its application in VLSI circuit design. Memetic Computing. https://doi.org/10.1007/s12293-020-00303-8
9. Liu C, Xu X, Hu D (2014) Multiobjective reinforcement learning: a comprehensive overview. IEEE Trans Syst Man Cybern Syst 45(3):385–398
10. Hourfar F, Bidgoly HJ, Moshiri B, Salahshoor K, Elkamel A (2019) A reinforcement learning approach for waterflooding optimization in petroleum reservoirs. Eng Appl Artif Intell 77:98–116

11. Li W, Özcan E, John R (2017) A learning automata-based multiobjective hyper-heuristic. IEEE Trans Evol Comput 23(1):59–73
12. Tozer B, Mazzuchi T, Sarkani S (2017) Many-objective stochastic path finding using reinforcement learning. Expert Syst Appl 72:371–382
13. Sun J, Zhang H, Zhou A, Zhang Q, Zhang K (2019) A new learning-based adaptive multi-objective evolutionary algorithm. Swarm Evol Comput 44:304–319
14. Chen G, Li J (2019) A diversity ranking based evolutionary algorithm for multi-objective and many-objective optimization. Swarm Evol Comput 48:274–287
15. Zhao H, Zhang C (2020) An online-learning-based evolutionary many-objective algorithm. Inf Sci 509:1–21
16. Vinay Kumar SB, Rao PV, Singh MK (2019) Optimal floor planning in VLSI using improved adaptive particle swarm optimization. Evol Intell. https://doi.org/10.1007/s12065-019-00256-z
17. Valenzuela CL, Wang PY (2000) A genetic algorithm for VLSI floorplanning. In: Parallel problem solving from nature PPSN VI, Lecture notes in computer science, vol 1917. Springer, pp 671–680
18. Cong J, Ma Y (2009) Thermal-aware 3D floorplan. In: Three dimensional integrated circuit design, part of the series integrated circuits and systems. Springer, pp 63–102
19. Potti S, Pothiraj S (2011) GPGPU implementation of parallel memetic algorithm for VLSI floorplanning problem. In: Trends in computer science, engineering and information technology, Communications in computer and information science, vol 204. Springer, pp 432–441
20. Liu J, Zhong W, Jiao L (2005) Moving block sequence and organizational evolutionary algorithm for general floorplanning. In: Computational intelligence and security, Lecture notes in computer science, vol 3801. Springer, pp 238–246
21. Tang M, Sebastian A (2005) A genetic algorithm for VLSI floorplanning using O-tree representation. In: Applications of evolutionary computing, Lecture notes in computer science, vol 3449. Springer, pp 215–224
22. Chiang C-W (2009) Ant colony optimization for VLSI floorplanning with clustering constraints. J Chin Inst Ind Eng 26(6):440–448
23. Deng YS, Maly WP (2010) Floorplanning for 2.5-D integration. In: 3-dimensional VLSI. Springer, pp 83–116
24. Aharoni M, Boni O, Freund A (2015) Rectangle placement for VLSI testing. In: Integration of AI and OR techniques in constraint programming, Lecture notes in computer science, vol 9075. Springer, pp 18–30
25. Singh D, Chandel R (2023) FPGA-based hardware-accelerated design of linear prediction analysis for real-time speech signal. Arabian J Sci Eng. https://doi.org/10.1007/s13369-023-07926-2
26. Chu WC (2003) Speech coding algorithms: foundation and evolution of standardized coders. Wiley, Hoboken
27. Skliarova I, Sklyarov V (2019) FPGA-based hardware accelerators. Springer International Publishing, Cham
28. Kumar S, Jha RK (2020) An FPGA-based design for a real-time image denoising using approximated fractional integrator. Multidimens Syst Signal Process 31:1317–1339. https://doi.org/10.1007/s11045-020-00709-0
29. Gersho A (1994) Advances in speech and audio compression. Proc IEEE 82:900–918. https://doi.org/10.1109/5.286194
30. Fenichel R, Bodson D (1992) Details to assist in implementation of Federal Standard 1016 CELP. National Communications System, Arlington
31. Xu J, Ariyaeeinia A, Sotudeh R (2005) Migrate Levinson-Durbin based linear predictive coding algorithm into FPGAS. In: 2005 12th IEEE International Conference on Electronics, Circuits and Systems. IEEE, pp 1–4
32. Menbari A, Jahanirad H (2022) A low-cost BIST design supporting offline and online tests. J Electron Testing. https://doi.org/10.1007/s10836-022-05986-0

33. Abramovici M, Breuer M, Friedman A (1990) Digital systems testing and testable design. Computer Science Press
34. Efanov DV, Sapozhnikov VV, Sapozhnikov VV (2017) Conditions for detecting a logical element fault in a combination device under concurrent checking based on Berger's code. Autom Remote Control 78(5):891–901
35. Floridia A, Mongano G, Piumatti D, Sanchez E (2019) Hybrid online self-test architecture for computational units on embedded processor cores. In: 2019 IEEE 22nd International Symposium on Design and Diagnostics of Electronic Circuits & Systems (DDECS). IEEE, pp 1–6
36. Jurj SL, Rotar R, Opritoiu F, Vladutiu M (2020) Online built-in self-test architecture for automated testing of a solar tracking equipment. In: Proceedings of IEEE International Conference on Environment and Electrical Engineering and IEEE Industrial and Commercial Power Systems Europe (EEEIC/I&CPS Europe). IEEE, pp 1–7
37. Wang C-H, Hsieh T-Y (2017) On the probability of detection lossless concurrent error detection based on implications. IEEE Trans Comput Aided Des Integr Circuits Syst 37(5):1090–1103
38. Jahanirad H, Karam H (2018) BIST-based online test approach for SRAM-based FPGAs. In: Proceedings of the Iranian Conference on Electrical Engineering (ICEE). IEEE, pp 178–183
39. Roth JP et al (1968) Diagnosis of automata failures. IBM J Res Dev 10(4):278–291
40. Divyapreethi B, Karthik T (2015) Input vector monitoring concurrent BIST architecture using modified SRAM cells. ARPN J Eng Appl Sci 10(9):4042–4046
41. Kochte MA, Zoellin CG, Wunderlich H-J (2010) Efficient concurrent self-test with partially specified patterns. J Electric Testing 26(5):581–594
42. Nikitha SA, Paulin S, Venkateshwaran SP (2015) A concurrent BIST architecture for online input vector monitoring. In: Proceedings of the International Conference on Science, Technology, and Management. IEEE, pp 1411–1488
43. Voyiatzis I, Efstathiou C (2013) Input vector monitoring concurrent BIST architecture using SRAM cells. IEEE Trans Very Large Scale Integr Syst 22(7):1625–1629
44. Voyiatzis I, Paschalis A, Gizopoulos D, Halatsis C, Makri FS, Hatzimihail M (2008) An input vector monitoring concurrent BIST architecture based on a pre-computed test set. IEEE Trans Comput 57(8):1012–1022
45. Pothiraj S, Kadambarajan JP, Kadarkarai P (2021) Floor planning of 3D IC design using hybrid multi-verse optimizer. Wireless Personal Communications. https://doi.org/10.1007/s11277-021-08166-z
46. Chiang T-Y, Souri SJ, Chui CO, Saraswat KC (2001) Thermal analysis of heterogeneous 3D ICs with various integration scenarios. In: International electron devices meeting, Technical digest (Cat. No. 01CH37224). IEEE, pp 31.2.1–31.2.4
47. Cong J, Wei J, Zhang Y (2004) A thermal-driven floorplanning algorithm for 3D ICs. In: Proceedings of the 2004 IEEE/ACM international conference on computer-aided design. IEEE, pp 306–313
48. Guo P-N, Cheng C-K, Yoshimura T (1999) An O-tree representation of non-slicing floorplan and its applications. In: Proceedings 1999 design automation conference (Cat. No. 99CH36361), pp 268–273
49. Sivaranjani P, Senthilkumar A (2015) 3D VLSI non-slicing floorplanning using modified corner list representation. Indian J Sci Technol 8:1
50. Sivaranjani P, Senthil Kumar A (2012) Thermal-aware non-slicing VLSI floorplanning using a smart decision-making PSO-GA based hybrid algorithm. Circuits Syst Signal Process. https://doi.org/10.1007/s00034-015-0020-x
51. Reddy SK (2015) Minimization of VLSI floorplan using hybrid cuckoo search and PSO. https://www.semanticscholar.org/paper/MINIMIZATION-OF-VLSI-FLOORPLAN-USING-HYBRID-CUCKOO-Reddy/a7a3fb0162bf4bc9b600f281c4f60ca50d9cba99

52. Kaur M, Kaur R, Singh N (2022) A novel hybrid of chimp with cuckoo search algorithm for the optimal designing of digital infinite impulse response filter using high-level synthesis. Soft Computing. https://doi.org/10.1007/s00500-022-07410-3
53. Aggarwal A, Rawat TK, Upadhyay DK (2016) Design of optimal digital fir filters using evolutionary and swarm optimization techniques. AEU Int J Electr Commun 70(4):373–385
54. Yadav S, Yadav R, Kumar A, Kumar M (2020) A novel approach for optimal design of digital fir filter using grasshopper optimization algorithm. ISA Trans 108:196–206
55. Wang Y, Li B, Chen Y (2011) Digital IIR filter design using multi-objective optimization evolutionary algorithm. Appl Soft Comput 11(2):1851–1857
56. Cuevas F, Gálvez J, Hinojosa S, Avalos O, Zaldívar D, Pérez-Cisneros M (2014) A comparison of evolutionary computation techniques for IIR model identification. J Appl Math 2014:1–15
57. Yang X, Deb S (2009) Cuckoo search via lévy flights. In: Proceedings of the World Congress on Nature and Biologically Inspired Computing (NABIC '09), Hindawi. IEEE, pp 210–214
58. Yang X (2010) Nature inspired metaheuristic algorithms. Luniver Press, Frome
59. Kim N, Xiong J, Hwu W (2017) heterogeneous computing meets near-memory acceleration and high-level synthesis in the post-moore era. IEEE Micro 37:10–18
60. Sengupta A, Bhadauria S, Mohanty SP (2016) Tl-hls: methodology for low cost hardware trojan security aware scheduling with optimal loop unrolling factor during high level synthesis. IEEE Trans Comput Aided Des Integ Circuits Syst 36(4):655–668
61. Geetha T, Sathya M (2012) Modified particle swarm optimization (MPSO) algorithm for web service selection (WSS) problem. In: 2012 International Conference on Data Science & Engineering (ICDSE) (12964092). IEEE, pp 1–8
62. Mirjalili S (2016) SCA: a sine cosine algorithm for solving optimization problems. In: Knowledge-based systems. Elsevier, pp 120–133
63. Bao G, Mao K (2009) Particle swarm optimization algorithm with asymmetric time varying acceleration coefficients. In: Proceedings of the 2009 IEEE International Conference on Robotics and Biomimetics (ROBIO). IEEE, pp 2134–2139
64. Russell E, Kennedy J (1995) A new optimizer using particle swarm theory. In: Proceedings of the Sixth International Symposium on Micro Machine and Human Science. IEEE, pp 39–43
65. Khishe M, Mosavi MR (2020) Chimp optimization algorithm. Expert Syst Appl 149:113338

MIX
Papier aus verantwortungsvollen Quellen
Paper from responsible sources
FSC® C105338

If you have any concerns about our products,
you can contact us on
ProductSafety@springernature.com

In case Publisher is established outside the EU,
the EU authorized representative is:
Springer Nature Customer Service Center GmbH
Europaplatz 3, 69115 Heidelberg, Germany

Printed by Libri Plureos GmbH
in Hamburg, Germany